POULTRY

A GUIDE TO MANAGEMENT

CAROL TWINCH

THE CROWOOD PRESS

First published in 1985 by
The Crowood Press Ltd
Ramsbury, Marlborough
Wiltshire SN8 2HR

Paperback edition 1993

This impression 2000

British Library Cataloguing-in-Publication Data

A catalogue record for this book is available from the British Library

ISBN 1 85223 755 4

Acknowledgements
The author acknowledges the help of the following: Michael Sheldrake, Betty Palmer, Derek Alsop, Oliver Illing B. Vet Med MRCVS, Harold Critchlow, Ann Knights, Tania Smith, Henrietta Buckwell, Paul Howell MEP, David C. Bland (Southern Pullet Rearers), Dr D.W.B. Sainsbury, Dr J. A. Hill (Gleadthorpe EHF).

Thanks for illustrations to SPR Poultry & Self Sufficiency Centre, Smiths Industries Ltd, Pintafen Ltd, Maywick (Hanningfield) Ltd, Curfew Incubators and Cobb Breeding Co Ltd.

Line illustrations by Annette Findlay (from sketches by the author).

Typeset by Inforum Ltd, Portsmouth
Printed in Great Britain at The Bath Press.

Contents

Introduction

Domestic hens were probably introduced into Britain during the first century BC. Although there is no firm evidence to suggest exactly when hens were first domesticated, two great prehistoric civilisations, those of the Indus Valley and the Yellow River Valley, are credited with doing much to help it along. It is fairly certain though, that all of today's breeds originated from four main varieties of Jungle Fowl – *Gallus ferruginus*, *Gallus stanleyii*, *Gallus sonneratii* and *Gallus furcatus*.

The four groups of *Gallus* developed independently in various parts of the world, very broadly grouped as Mediterranean, Asiatic, British and European. As the first two found their way into the other areas, breeders integrated them and crossed to new strains, arriving at what are now accepted as 'pure' breeds. These identifiable breeds ruled supreme until the commercialisation of poultry began at the end of the Second World War when genetic advances were made towards the introduction of prolific hybrids, both for egg production and the table, which were able to perform well under intensification.

Generally speaking, the varieties of fowl from the Mediterranean region were 'Light', that is to say they were excellent egg producers, with a body weight of less than 5½lb (2.5kg) plus a nervous and flighty temperament, not conducive to broodiness. The *Leghorn*, a breed developed in the United States from an Italian bird, is always thought of as a prime example of lightness. The distinctive black *Minorca*, with the white earlobes, comes a close second as a producer of large white eggs.

Another famous result of the skilful blending of European and North American breeds is the *Rhode Island Red*, the most common representative of the 'Heavy' breeds. Strains of *Shanghais*, *Malays* and *Javas* were crossed with *Brown Leghorns* to produce the world's best known fowl.

One of the oldest British breeds is the *Dorking*, which is thought to have been already established at the time of the Roman invasion. The Romans managed their fowls commercially, that is for food as well as sport, and recorded for history their dismay at finding the Britons kept their fowls mostly for the sport of cock-fighting.

Today, there is a revived interest in pure breeds, exhibiting and backyard chicken keeping. For a variety of reasons, all kinds of people are being attracted to small scale poultry keeping. In some cases, the skills of economical egg production have to be learnt from scratch, while in others a new interest beckons in showing, an expanding and increasingly absorbing hobby.

This book hopes to include something of interest to the old hand and the total newcomer. Usually the basic husbandry needs are adapted to the individual limits of space, time and money. The fact that the opinion 'there are as many systems of keeping poultry as there are poultry keepers' has been said many times before, makes it none the less true. Nearly all the ideas and suggestions mentioned must be adapted to the best personal advantage. There would be little point in embarking upon traditional-style chicken keeping unless it provided satisfaction and enjoyment as well as good, fresh eggs.

During the writing it became increasingly difficult to find another word for 'chicken'. A purist poultry keeper pointed out that strictly speaking the word should apply only to male or female chicks until their first birthday. But the term is so commonly used to embrace all ages and types of Gallinaceous fowl that it now has an accepted usage.

1 Breeds

Decisions about the type of birds best suited to individual needs are best made after a close look at the alternatives. Some breeds are primarily egg layers, others table birds, and many more all-rounders or dual purpose. The five main reasons for keeping chickens are:

Egg production
Meat production
Dual egg and meat production
Maternal instincts (i.e. broodies)
Showing and pedigree breeding

To the beginner, there isn't much difference between one of the commercial egg layers and a pure breed. The prejudice against 'old fashioned' types of poultry has been encouraged by hybrid breeders, who often think that no other birds can lay eggs. This, of course, is completely untrue. The concept of backyard egg production is entirely different from a commercial enterprise, and the choice of breed has to be made according to needs. While any commercial hybrid will do very nicely in almost any husbandry situation, except of course a bad one, this is not always the case with pure breeds.

PURE BREEDS

There are about fifty breeds listed in the British Poultry Standards, not including Bantams, which number almost as many again. A short description of some of the most common types will help the novice identify these, and provide some useful background information.

Some of the rarer breeds can be difficult to track down when it comes to buying in pullets or hatching eggs. It is, however, less difficult now than it has been in the past, and an inquiry to the Poultry Club or Rare

Breeds Society will usually bring some positive response. There has been an increase lately in specialist breeders, and therefore a greater range of breeds is steadily becoming available for purchase.

It is sometimes suggested that a beginner should start off with some guaranteed, point of lay 'brown' pullets, with which, apparently, no one can go wrong. This is not the case. Anyone can go as wrong with six Marans as they can with six Golden Comets, and none of them will thrive if they are not fed and looked after properly. Here are a few brief details about some of the most common pure breeds.

Leghorn

Of Mediterranean origin and classified Light. There are several varieties in colour, but the most common are black, brown and white. The White Leghorn in particular has featured in several of the high egg laying hybrids. Considered by some as 'flighty' this can as easily be called 'alertness'. Unlikely to go broody. Currently marketed commercially in its pure form, and it is claimed to lay up to 280 eggs per bird to 74 weeks of age.

Maran

Of French origin, and classified Heavy. This breed has enjoyed a considerable revival during the last few years, mainly because of the very deep rich brown colour of their eggs. The black and white barred Cuckoo Maran is the most common variety, and is more of a general purpose bird than the Leghorn. Although not top in the egg producing stakes, it will provide good meat. Maran eggs can sometimes cause problems with hatching, as the shell is very hard. The hens make excellent mothers.

Rhode Island Red

Easily the most famous breed in the world, its origins are in North America. Classification, Heavy. It takes its name from the farms of Rhode Island Province where it was formed and since it was standardised in 1909, has been the most successful of dual purpose birds. Egg colour is light to medium brown, and the females good sitters. The

Commercial Rhode Island Reds on range.

cockerels can be aggressive. Easily obtainable. Suppliers now claim that commercially bred Rhodes can produce 300 eggs per bird in a 12 month laying period and will lay continuously for up to 15 months.

Silkie

Of Asiatic origin, classified as Light but has a persistent broodiness as a breed characteristic. It is often referred to as a Bantam, but it is exhibited as a large fowl. The white variety is the most common.

Sussex

A British breed, classified Heavy. Probably second to the Rhode Island Red in popularity and usefulness. Of all the varieties, the Light is the best known. At the time when sex-linkage was first discovered, the two breeds were in considerable demand as together they produced the required offspring, i.e. chicks which could be identified at birth as male or female by their down colour. Considered a very fine table bird, but nevertheless a good layer. Not so inclined to broodiness as some other heavy breeds, but certainly reasonable mothers when they do hatch.

Welsummer

A European breed of Dutch origin, and classified Light. Their egg colour is a beautiful deep brown, though not with quite the depth of the Maran egg. Unlikely to go broody. Good layers and although, like the Leghorn, Minorca and other light breeds, thought to be flighty, they do respond to handling.

COMMERCIAL HYBRIDS

The advantage of buying commercial hybrids is that they are bred for the job, whether egg producing or putting on meat. They are uniform in their growth capacity, maturity and performance, although their producers' claims for maximum egg production must be looked at in the light of the conditions under which they are kept.

The old arguments of white versus brown eggs somehow continue into a scientific age. There can be absolutely no difference between a white egg and a brown one, except psychological. For some reason the Americans prefer white eggs and the British like them brown. The commercial breeders have therefore done their best with both types. The white egg layers are the Hubbard White Leghorn and the Shaver Starcross numbers 444 and 288. Shaver also have two brown egg layers, the 566 and 585. Other hybrid brown egg layers are: Hubbard Golden Comet, Ross Brown, Ross Ranger, Tetra SL, the ISA Brown and Babcock B.380.

Their point of lay is almost always between 18 to 20 weeks of age, with a body weight at that time between 3.2 and 3.4lbs (1.45–1.55kg). Body weights in hybrids do not indicate lightness or heaviness, as in pure breeds. They are usually between 4.5 and 4.7lbs (2.05–2.15kg) at maturity.

Average feed consumption figures are always obtainable from the companies, but they must be adjusted to account for non-typical housing and environment. Nearly all the feed conversion and liveweight gain figures are dependent on conditions being similar to those of commercial rearers and producers. For those companies, performance records are indispensable for publicity purposes and usually follow along the lines of the detailed advice for commercial producers which is contained in the Ministry of Agriculture leaflet No 599. A lot of information is contained in the leaflet though some will be superfluous to the needs of a small scale producer. Various flock costing schemes are run by the Ministry, and results from tests carried out at the Experimental Husbandry Farm at Gleadthorpe are available in booklet form. The performance of particular hybrids in the egg laying and weight gain recordings are the results of the most meticulous record keeping, thus enabling comparisons to be made which may influence the choice of purchase.

The commercial companies don't do things by halves. ISA, for example, market around six million day-old chicks in the Eastern Counties alone, as well as supplying 300,000 parent stock to UK franchise hatcheries. It might, though, sometimes be difficult to track down just a dozen or two of these six million, as hatcheries do not usually deal in low number orders. However, they will direct such enquiries to a contract rearer.

BANTAMS

Bantams are no more and no less than miniature versions of the larger fowl. That is not to say that there are no true Bantams, but that is a matter of academic importance. Most of the larger breeds have Bantam equivalents which have been bred over the years, using small pullets and selecting scientifically for smallness.

Bantams are particularly popular with owners of small gardens or limited space. Less housing is required and their upkeep costs proportionately less. Showbirds need smaller pens and carrying baskets, again reducing costs. Although Bantams eat less food than larger fowl, they still need to maintain body condition as well as produce eggs – both proportionate to their size. A Bantam's crop size can be compared with a small table tennis ball and holds about 1oz (28g) of food when full.

Because, for these obvious reasons of cost and space, Bantams have been popular with the Fancy, there is a wide variety of colours, types and ornamental birds, specially bred over the years by dedicated men and women with a love of chickens.

The collective term *Bantam* is supposed to have been derived from a town of that name in Java. The story goes that crews on the trading ships used to operate a black market in birds and animals bought on their travels, and it was at Bantam that some particularly small fowl could be found and smuggled on board. No one breed though has been traced back to a particular town named Bantam.

The breed characteristics for Bantams are much the same as for their larger equivalents, though particular strains will differ, much as in large fowl breeds. Not all Bantams are identical to their larger counterparts, for example, the Bantam Welsummer eggs are not as noted for their deep rich brown as those from the large fowl.

Bantams are not so amenable to winter egg production or enforced moults, and are far closer to the natural cycles of their jungle fowl ancestors.

TABLE BIRDS

Some breeds of chicken have been selectively bred for the purposes of quick growth and plenty of succulent meat. Before the development of hybrid broilers, the Light Sussex, in spite of its comparatively slow growth, was the popular choice for those with egg production as a second option.

Over the years, various methods of quicker fattening have been tried, the most popular until a few years ago being caponising. This involved an operation to remove the male reproductive organs at eight weeks of age, and later the implantation of chemical hormones into the neck of the birds. Capons are mentioned in history as far back as Roman times, but what their definition was is hard to discover.

Both caponising methods had the effect of removing the male characteristics of aggression, depressing the size of the comb and their ability to crow, and generally rendering them docile, thus encouraging quicker growth, and increasing the final weight. Both practices are now illegal, thus giving a boost to the selective breeding of broilers which can reach a killing weight in about 12 weeks. The extra labour needed to caponise large numbers of birds also ensured the success of the broiler.

Tales of fat, lazy capons being 'trained' to brood young chicks have appeared in poultry books belonging to the last century. Experiments over the years into the effects of male and female hormones in chickens have found that eggs are mostly produced from the female's left ovary. Should that ovary cease to function, either from disease or age (over 10 years), the female bird can take on male characteristics, even to the extent of sporting male plumage. Few birds these days, however, would be permitted to get to that stage in their evolution.

The broiler chicks are sold as male or female, or as hatched, and although the male chicks will reach their killing weight slightly quicker than the females, this might not be a bad arrangement for the small scale producer. The same companies which have bred the successful egg layers also produce the broilers. Ross, Cobb and Hubbard are the best known and although these companies are now multi-national and therefore not involved in direct selling, their local agents or contract rears might help with the problem of availability. ISA Poultry Services

The Cobb Breeding Co offers a range of broiler birds.

have added the ISA Vedette broiler stock to their two existing brown egg layers.

The performance of the Ross Cobb 500 gives a guide to broiler performance, provided that the correct food is given and they are not range reared. At 35 days of age the male Cobb will weigh approximately 3.79lb (1.7kg) and at 49 days of age will be 5.75lb (2.6kg).

2 Stock Selection

Having decided what type of birds to buy, and for what purpose they are required, there is now the bewildering task of selecting a supplier. Not so long ago, it was difficult for small scalers to buy six or a dozen pullets, but under the law of supply and demand, there are now any number of commercial stockists, not only for pullets, but for day-old chicks, or mature hens. As well as supplying stock, most of them have the necessary paraphernalia on sale, either on site or by post. A look through any of the poultry or self-sufficiency type magazines will usually provide a good selection. Most of the general farming magazines also have advertisements for suppliers and breeders.

Unless you live within driving distance of the supplier (and most of them will be happy to let small numbers be picked up) the birds will be sent by rail. One or two suppliers have their own transport and will make free delivery on two or three days a week within a stated radius, but generally stock will be sent to the nearest railway station. There are several A4 pages of rules and regulations governing the movement of stock by rail, drawn up by the British Small Animal Veterinary Association, in consultation with the Medical Research Council Laboratory Animal Centre, the RSPCA, and British Rail. However, the most important criterion is that containers should be rigid and allow the birds some movement and plenty of air. A very reasonable regulation, and one which is enforced by spot checks and attentive rail staff.

POINT OF LAY

There is no best time to buy in the first lot of stock, but the traditional seasons for Point of Lay (POL) pullets are spring and autumn. Their age will be about 18 weeks. For the birds to begin laying in March or April, they will have been hatched in the previous October, while the March or April hatchings will begin their laying period in September.

Some of the lighter breeds can be hatched as late as April, but the heavier general purpose birds need to be hatched at the beginning of March. Hatching time is important. Too early and they will lay a few eggs and then moult, while too late means that they may not have reached sufficient maturity to cope with the demands of 12 months continual egg laying.

Some backyarders find it convenient to have six autumn hatched and six spring hatched, with the aim of having year-round eggs.

To some extent the laws of moult will depend on the type or breed of chicken, but there is no reason why an enforced moult cannot be carried out to ensure the maximum laying potential. Notes on winter lighting and moulting can be found in Chapter 5. If the pullets have been home reared it may be necessary to stop them laying during August or September if they are required to go through to the next spring.

Buying in stock always means that only part of their egg laying potential is in your hands, though with the commercial hybrids they have usually been reared to the maximum efficiency. A good choice of parents and a high standard of rearing are both things of the past. Breeding 'will out' and no amount of management can overcome poor stock. Alternatively, a highly bred fowl can often give reasonable results under dubious conditions. With some of the pure breeds, it may still be a question of quantity not quality for the commercial producers. What the purchaser gets in the way of stock in, say, the *Marsh Daisy* or the *Scots Dumpy*, is something like roulette.

The way of overcoming this problem is to buy birds with a more predictable record. There used to be a feeling that the commercials did not do so well in backyard situations as the older breeds. That could well have been the case with the prototype hybrids, but it is not the case now. It is, however, a valid observation that higher results are got from battery birds because they are 'molly coddled' and not because they are necessarily better birds. It is a question of what is wanted. There are plenty of cheap, battery produced eggs in the shops.

Pullets of 18 weeks of age will usually need a week or so to settle in. Sometimes they will be unused to perching facilities, and they can be helped by lifting them onto the perches after dark. If the pullets are actually on the point of laying, or have already started to lay, nest boxes must be provided immediately. Once they begin to lay on the house floor it might be difficult to stop them doing it consistently, thus causing untold problems for the future in lost eggs and egg eating.

DAY-OLD CHICKS

The buying in of day-olds brings more control of their productive life, but it does mean that unless someone can lend you a broody hen, a lot more equipment will be required. It is also possible that in buying twelve 'as hatched' day-olds, there could be only three or four pullets among them. 'As hatched' is the common way of buying pure breeds, though of course the commercial types are sold as pullets. There is a greater danger of chilling in makeshift brooding facilities, which are likely to be the type offered. It just is not economical to invest in expensive brooding equipment for small numbers of chicks and the variations on the theme of hay boxes often leaves a lot to be desired. The mortality rate for day-old chicks is also higher than for POL pullets and they will need greater protection from predators.

MATURE HENS

Occasionally, a beginner may be offered a few hens from a well-meaning friend, or someone whose breeding methods have been too successful for their own requirements. In other words, stock which is looking for a good home.

Although a bad experience is often worth two good ones, these birds are not the best buy, even if they are free. Sometimes they are in early moult, which means keeping them for an indefinite period, but more often than not they are just not 'doers'. If Lady Luck is running really low in her largesse the stock can be vice-ridden, either egg eaters or aggressive to other birds. A friend of course would not pass on such stock, but there are varying degrees of friendship, and a novice would be best advised to study the ways of assessing the worth of such poultry before taking it home. It is also useful knowledge to have if a mature laying trio is being purchased for specialist breeding, although in that case the buyer will already have quite a good idea of what he or she is looking for.

One of the best ways of gauging the age of the fowl is to study the leg scales. They should be compact, lying neat and flat. The younger the bird the softer the scales. The same goes for the face detail – which should be smooth and soft to touch. If these two things seem all right, the next check is with the wing feather. Details of the moult are given in Chapter 5, but briefly the last part of the full moult is in the wing. In light coloured breeds it is easy to spot the new, clean feathers and if at least half the wing has been renewed the bird is well on her way through. The other well-known test is the fingers test. This should indicate where the pelvic bones are. Roughly, they are an inch or two above the top of the legs and just below the vent feathers. The sign of a good layer is the width of three fingers. If only two fingers fit between the two bones the hen is probably in lay, while one finger means it is not.

Even assuming that you have average sized fingers, and the hen is also of average size, this test is a guide only and should not be taken as absolute proof of the state of the bird. It is too easy for an amateur to go wrong. Hens, like people, differ.

A breeding trio should carry with them some degree of comeback if either the hens or the male develops a disease or mechanical problem. It is better to buy from a reputable breeder, after visiting the establishment, and where possible some written evidence of the transaction. It is not satisfactory to purchase a breeding trio on the way home from the seaside on impulse from the roadside farm.

Most responsible vendors will be happy to give an assurance that the stock has been blood tested in order that any carriers of Marek's disease can be identified. More details on this can be found in Chapter 8, but basically, all bought in stock should have been vaccinated against the main blood transmitted diseases.

BATTERY REFUGEES

Some success can be had with second year battery hens. Such birds have usually gone through their peak laying period and their performance for the second term can only be uneconomic for the commercial enterprise. Strangely enough, it is often these birds which are inclined to be friendly towards humans, presumably having been tamed by cage life and well used to being handled. After their initial de-batterification, they often make ideal free-rangers, and are not so inclined to 'lay away' as birds reared more naturally.

Battery hens will often be much cheaper than pullets but they do need time to recover from life as an intensive egg producer. Their appearance is often shocking to anyone expecting to see plump, majestic birds; they usually have few feathers, not necessarily because they have been feather pecking but because they are in moult, and sometimes have difficulty in walking. However, their recovery and normalisation is very rapid and most satisfying to their owners. Check that their beaks have not been cut back too severely, that they are able to peck up unfamiliar food and that water is easy for them to find.

COCKS AND COCKERELS

There is no need at all for anyone keeping a few chickens for eggs or meat to keep a cockerel. Unless required for breeding, he can only be a nuisance, except on free range and under idyllic conditions, perhaps similar to the old fashioned farmyard.

Males for breeding must be selected carefully, but whether high class or mongrel, he will still greet each new morning with traditional crowing. It is as well to check that no Local Authority by-laws or clauses in house purchase contracts, tenancies or leases, exist which forbid the keeping of poultry. Complaints to the Department of the Environment, which deals with such things as noisy cockerels, are very common, and usually the complainant wins the day.

3 Housing and Equipment

'There are as many systems of housing poultry as there are poultry keepers . . .' The fact that this has been said over and over again makes it no less true. So much depends on home resources, space available and money to spend. However, the principles of good husbandry are easily adapted to individual needs and guidelines for the basic requirements of stock have endured fashions and trends in all kinds of chicken enterprises.

HOUSING

The amount of space available for the housing has, to a large extent, influenced decisions on the type of poultry most suited to the conditions. In some ways, the less space there is, the easier the choice.

As housing is a considerable part of capital outlay for any system of poultry keeping, careful thought has to be given to using cheap, easily available or existing opportunities.

In a farmyard situation this is easy. Plenty of yards are still the playgrounds for the free-ranging collections of miscellaneous hens, who scratch a living and produce a seasonal crop, both of eggs and offspring. The first though must be found – the second, caught. Not very scientific, but certainly cheap and 'natural'. Costs in this situation are zero and anyone keeping chickens under these circumstances is not in need of much advice, unless they want to improve output and tidy up the system.

Where show birds are kept, special arrangements need to be made for pens and show training. More details on these requirements are outlined in Chapter 9.

Broadly, the two types of housing are *permanent* and *movable*. The first is usually a converted building, or part of one, adapted to specific needs. Alternatively, a shed too large (or too old) to be moved. The second is a hen house, small and sturdy enough to be moved at intervals, with or without an enclosed run attached.

Permanent Housing

There is almost no building which could not be made into suitable accommodation for either laying fowl or fatteners. Even the most dilapidated abode can be made rat proof, by filling in wall and floor holes with balls of wire netting or broken glass and sealing with concrete. So long as repairs can ensure weatherproof roofing and intruder proof walls, plus an even, cleanable floor, it can house poultry. If it is not possible to add a straw yard or run, a system of semi-range can be operated, as outlined in Chapter 4. Alternatively, a deep litter system can make full use of any capital employed. A small stable of about 8ft × 16ft (2.4m × 4.8m) will house up to 25 birds. The capacity of any building can be worked out by allowing 4 –5 sq ft (.37m^2–.46m^2) per bird. If a yard or run is available, this can be reduced to 3 sq ft (.28m^2) per bird but it is always better to over-estimate the space required than crowd in too many.

If there is to be no access to the outside, more thought must go into the design of the interior. A fitted frame of wire placed at the door opening with a removable piece of wood at the bottom to contain the litter would allow some sunlight and plenty of air into the interior.

FLOORING AND WALLS

To do the job properly, the floors and walls should be free of holes, especially if built of the old lathe and plaster. The hens will gradually peck away at any small openings and it is far easier to get it right now – at the beginning – than half way through the first laying term.

Repairs and renovations must be done when the shed is empty and they are not going to scare the living daylights out of the inhabitants. A few airflow bricks can be incorporated into the walls if repairs are severe, but not too low down as they can interfere with the effectiveness of deep litter and will get bunged up with straw and muck. Too far down would be less than 3ft (90cm). They should also not be placed opposite one another, thereby minimising the draught. A cheap alternative to air bricks used to be open drainpipes built into the brickwork, but this has the obvious disadvantage of allowing vermin to enter the shed.

Concrete is the best material to use for making a smooth surface, and one which can easily be hosed down between batches and requires no maintenance. If the inside is dirty, an initial pressure hosing will be required. A pressure washer can be hired for a day or two. Concrete is also not so prone to harbouring pests and any attempted entry by rats can easily be spotted. Except when it is being hosed, the flooring will not be required to handle the amount of water normally associated with indoor livestock, but it is helpful to make a slight slope towards a drain in the place where the water will be situated. If the drinking system consists of a tap and bucket, which has to be swilled out daily, the swilling should take place at a drainage point, not washed out onto the litter.

Adapted housing with convenient food storage outside.

Strong weatherproof housing, built to last. Suitable for up to fifty layers.

ROOFING AND CEILINGS

It is probable that the shed is at least waterproof, or it could be with the rearrangement of some slates, new roofing felt or sheeting. Weather-proofing is most important, and leaks merely undermine the good work done in making the flooring decent.

If the latest range of building materials, and methods of DIY maintenance and repair, has not previously been of interest it is a good idea to purchase one of the many DIY magazines. Two or three of these will usually give a newcomer to the building world some idea of what is possible and at what price.

Birds, mice and rats will soon make use of any access points where the walls meet the roof, so these again must be sealed, though not with any materials which will hinder the free air flow. If the ceiling is very high, a false roof can be constructed using wooden stays, wire netting and straw. Insulating boards can also be used, though both can be a fire hazard. A few overhead beams can be most useful to hang feeders and grit dispensers on, or for wiring if electricity is possible. However, false roofs do collect dust and must be carefully constructed to be able to sustain the weight.

VENTILATION

The free movement of air is particularly important if the building is to house the birds for a large part, or all of the time.

The risk of respiratory infection is increased dramatically by moist, stale air. Air coming in through vents or windows must have somewhere to go, hence the need for some outlet at roof level. If the air circulation is to rely on a large opening, such as a window, it should be remembered that good ventilation is not just air movement any cost.

A draught can be construed as providing circulating air, but as a faster rate than is desirable. To combat quick intakes of outside air (and blasts of stuffy summer warmth can be as damaging to the enclosed environment as icy winter cold), the opening can be baffled on the outside. This prevents air being brought directly into the building, making it turn corners, thus slowing it down before it enters the interior. It also prevents driving rain from entering.

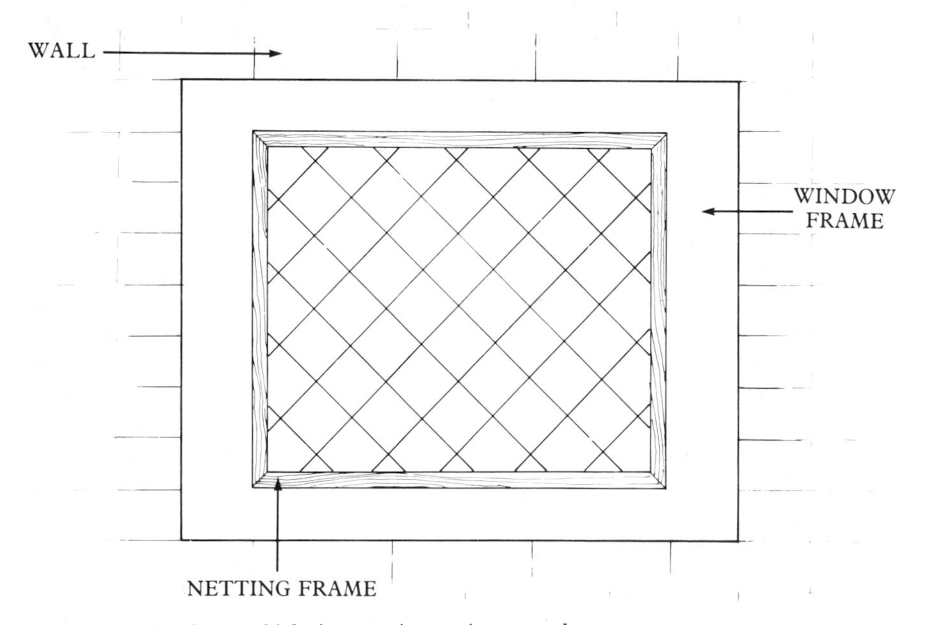

Fig 1 Netting frame which gives maximum air passage but prevents escape. If black polythene is placed behind the frame, it can also be used to exclude light.

DOORS, WINDOWS AND POP HOLES

In an existing building, there is little that can be done in the way of design, except to make efficient use of what there is. As already suggested, a netting frame can be used in a doorway, and similarly in windows which can be either fixed or removable.

If the latter, frames must fit well and not be easily blown in or moved. If more control is required at any time over the amount of daylight reaching the birds, the frame can be used with a sheet of black polythene, cut to several inches larger than the outside rim of the wood, and pushed into place. This might be necessary if an exacting winter lighting pattern is required. If the window was previously part of the ventilation system, compensation should be made.

If the shed or building has, or can have, a scratching yard or run for the birds, a pop hole will be required. This must not be a weak spot in the security system. When closed it must be as impenetrable as the other openings, and when open must allow comfortable passage to and from the housing. If it is placed near to the edge of the run it can easily be operated from outside. If closing it at nights requires little effort, it will more likely be done!

ELECTRICITY AND WATER SUPPLY

If electricity is required, and it is not already available, the wiring is a job for a professional electrician. Recommendations for the choice of fire resistant materials are outlined in the Ministry of Agriculture *Code of Recommendations for the Welfare of Livestock*. Although a lot of the recommendations are designed for commercial producers, it is nevertheless a guide to the standards of safe stockmanship required by law.

While pricing up the cost of installation it is worthwhile putting in a light timing mechanism and a dimmer on the switch. The latter gives a warning to the birds that it is bedtime when a winter lighting pattern is in operation.

If a light is already in situ, but not centrally located, all that is required is extra wiring to suspend it well above the level where any bird could reach it. A plain white shade is all that is needed to spread the light evenly. If starting from scratch, strip lighting can be installed, which is cheaper to run. Depending on the shape of the building it may give better light distribution.

The same principles of convenience apply to the water supply. If it is relatively simple to route a water pipe into the shed, it is worth having a drinker installed. There are several different kinds available, but when locating the wall mounted ones remember that the level of the litter will build up. To begin with, a brick can be placed in front of it to allow access. Free-standing drinkers are not recommended for deep litter, but many drinkers of that style can be suspended, if roof timbers allow. Some of the older, heavier drinkers, filled with two gallons of water are heavy, though the new modern plastic ones would be better suited.

Some of the automatic drinkers can be connected directly into a

Plastic drinkers, suitable for free standing or suspended use.

½ inch (12mm) or ¾ inch (18mm) galvanised or plastic water pipe. If the pipes are running outside they need lagging. The foam rubber tubing which is split to fit round the pipe is the best and most convenient way of protecting the bird's water supply. The tubing is held with bands of sticky tape and will flex round bends and cut to irregular shaped corners. For extra protection this in turn can be wrapped over with sacking and made waterproof with polythene. The foam rubber by itself will not stand up to winter weather and once wet will be worse than nothing.

OUTSIDE SCRATCHING AREA

In ideal circumstances, it will be possible to extend the area of housing to include access to the great outdoors. Practically, this should be enclosed, escape proof, or at least with high enough netting sides to discourage escape attempts. Clipping the fowl's wings will help with that problem. The cuts must not be too severe as there are blood vessels in the lower half of the feather. Most birds can be confined without clipping, but some Bantams or flighty light breeds may need doing.

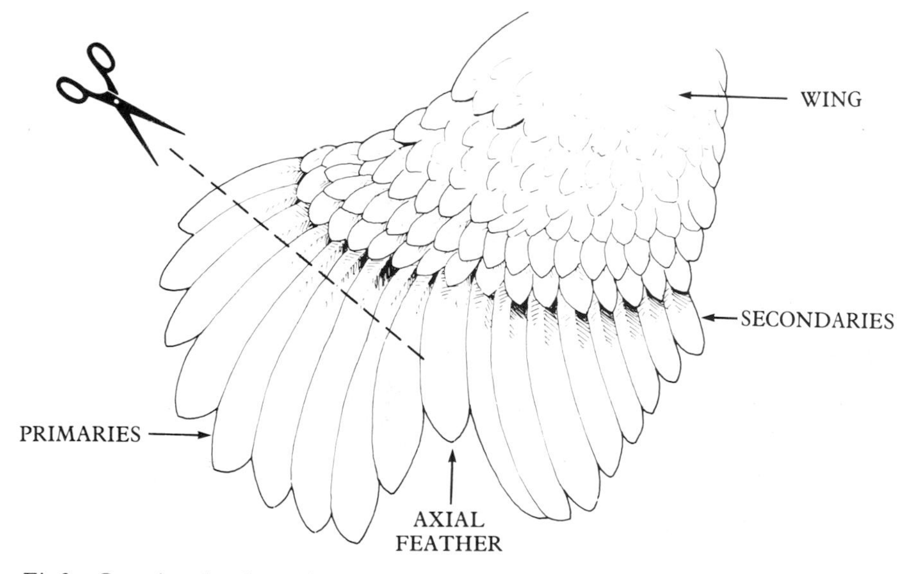

Fig 2 Cut primaries of one wing only, being careful not to sever quill blood vessels.

Only one wing is done, to cause imbalance, and needs to be done after the moult if the bird is kept for more than one season. Show birds naturally are restrained in other ways. Generally, the more area the hens have to roam around in, the lower the perimeter fencing needs to be, though 5–6ft (1.5–1.8m) would keep most poultry in.

Metal wire netting is still the most common material to use, but there is now a netting made from knitted polyethylene which is claimed by its manufacturers to be '. . .tough, durable, reusable, rust free, weatherproof and easy to handle – not to be confused with extruded plastic netting which suffers from the weather and is usually difficult to handle'.

Whichever material is used, it is essential that it is escape proof. If second-hand wire netting is being used, it needs checking to see whether there are any weak spots. Once a hole appears and has to be patched, it will give a predator an opportunity to enter the run and the birds an escape route. It is not much fun to spend days putting up netting, or making a run, only to find it is not effective and needs doing again.

If the run area is small, it will not be practical to think it will provide much in the way of feed, i.e. grass, unless definite steps are taken to ensure its management. The bare patch of earth will soon be unsightly and unhealthy. It can therefore be dug over and covered with gravel or stones after the day or two it takes the fowls to clear it. In dry weather, some loose straw can provide lots of scratching materials. Although unsightly, about the only things which will flourish in a hen yard are bracken and nettles. Although cutting them down is tidier, the hens do like to scratch and dust bath beneath their shade. Presumably it reminds them of their ancestors' days in the jungle.

Movable Housing

Movable housing can be divided into two broad divisions, *semi-permanent* and *portable*. Both are usually made from wood, although there is now a plastic coop on the market. The very small, all-in-one houses come into this category.

The first is usually one which is capable of being moved, not necessarily used in a formal folding system, but often left on range for long periods without changing position. The second is often smaller

Movable housing can be left on range for long periods with low stocking rates.

Ideal movable housing for 6–8 layers.

and usually has a run attached which is moved regularly round a garden or field.

If looked at purely from a numbers point of view, the amount of ready-made, expertly constructed wooden chicken houses is daunting. However, if bought brand new, the realities of space and money will probably narrow the field considerably. If it is a matter of renovating an existing house, the design is probably not dissimilar to those on today's market, and need only be made weather and vermin proof.

Give it a really good clean inside and out, and do the carpentry where necessary. For preserving wooden components, creosote can be used but it has some disadvantages. It has to be applied every year, it has no ability to colour wood anything other than a dull black which bleaches in the summer, and there can be a danger to the stock if they are put into the housing too soon after its use. It can also contaminate grass and plants. However, one of its advantages is that it can be used on damp wood (unlike a preservative which would seal the moisture in). It also discourages insect life from colonising too quickly in the wood's crevices.

If a degree of disinfection is required (over and above the need to preserve the wood), for example on the inside for perches and nest boxes, creosote does the job well. A good ten days to a fortnight should be allowed before birds are put into it, especially if they are to be enclosed at night, to allow the fumes to die down and the liquid to be absorbed into the wood. Special attention should be paid to perches and corners which can harbour pests. There is no substitute however, for thorough cleaning, so any existing muck or straw must be scraped off before treating.

Ordinary paint is not a good idea, as it will wear and flake, encouraging the birds to peck at it and make the house and run look a mess.

When it comes to starting from nothing, the choice is between ready-made and home-made houses. There are plenty of suppliers around, all of whom offer a choice to suit all needs. Most of them carry a recommendation for the maximum number of birds which they will house.

DIY

Starting from the beginning can have its advantages, the chief one being

Fold unit suitable for rearing young stock.

Semi-permanent fold unit with plenty of head room in the run.

that the housing can be tailor-made to fit into the proposed enterprise. Provided that care is taken with the purchase of materials, a second advantage will be the lower cost. Here are some points to bear in mind when designing and building a hen house.

Simplicity Keep the design simple. This will reduce both the initial cost and future maintenance.

Accessibility Make sure that all parts of the house can be cleaned and disinfected adequately. It is also practical to reduce day to day work to a minimum. For example, placing the roosting perches above a droppings pit or board is decidedly more practical than siting them above the nest boxes or grit trays.

Intruder proof Take all precautions against the intrusion of enemies, whether rats, foxes, dogs, cats or anything else which could frighten or kill the hens. They can die as easily from fright as actual harm. All doors and pop holes must be able to close.

Weatherproof Important to both the health of the chickens and the longevity of the construction. A cold, wet and draughty hen house is no good to anyone.

Economy While keeping costs in mind, think of the construction as a long term investment. False economy is out, but economy is definitely in.

Look around Have a good look at as many differently designed houses as possible. Think about the number of fowls it will hold and the system of husbandry required. A fold for rearers needs a much simpler design than one required for layers. A cleverly designed house could be used for both, with the minimum of rearrangement.

POINTS TO CONSIDER

Flooring Solid floors are more expensive than wire ones, but if the house is to be on range it will need to be substantial. Alternatively, a wire one at earth level gives easier cleaning and access to grass for the birds. It is also safer for broodies and their small chicks or young chickens after weaning. Their legs are easily broken and can get caught in slats, which can also be chewed through by a hungry rat.

Walls and roofing These need to be considered in the light of where the perches and nest boxes are to be. Perches are heavy when eight or ten birds are on them. Materials used must be weather and vermin proof as well as strong enough to carry the weight of any water or feed dispensers which will hang inside.

The method of moving the house must be calculated in the initial planning stages. Constant movement will undermine any weak points in the construction, and whether on wheels or sedan-like holding poles, the floors, walls and roof must hold together. Wooden ski bottoms can often put less wear and tear on the construction.

Free standing nest boxes suitable for up to ten birds.

Windows and pop holes must fit securely. If part of the walling or ventilation is to be wire netting, baffles, similar to those described for permanent housing, can be fitted to prevent rain or wind from entering the interior. Any wire netting should be as new as possible, free from holes and small gauge to keep sparrows and their like out.

If the housing is to be used for young stock, or rearers, there is less need for perches. Both these and nest boxes therefore can be removable. This not only makes them easier to keep clean but allows more birds per square foot. Without perches, floor level draughts are inadmissible. Perches for layers will be required at a length of 6–8 inches (15–20cm) per bird and be strong enough to take the weight of however many they are designed to hold.

Nest boxes can be fixed to the wall, or free standing. Generally, one nest compartment should be allowed for every three birds. Lack of adequate nests will result in dirty or broken eggs, and unsuitable ones will encourage the birds to lay away if on range.

Windows will not only allow light in, but help with air movement. A piece of glass or polythene can be fixed over the netting opening if a baffle is not possible. If it is made so that it can be slid backwards and forwards across the windows it can reduce the problem of winter weather reaching the occupants.

Roofing need only be simple corrugated sheeting or weatherproof wood. One section of corrugated roofing could consist of clear plastic, to let in plenty of light. In summer however, it can also let in too much sun, but this can be overcome by using some greenhouse shading on the inside.

FIXTURES AND FITTINGS

Before final plans are made, consider where the hens will enter and leave the house. The entrance need not be very big, 1ft (30cm) high will be plenty, but if a permanent run is to be fixed it needs to be opened and shut easily from outside the run.

A side-sliding door is often more practical than an up and down one and is easier to rig. There is less chance of it shutting by mistake and shutting the birds either in or out.

The height of the run should be sufficient to allow birds to stretch and flap their wings, especially if a cockerel is to be housed. Generally,

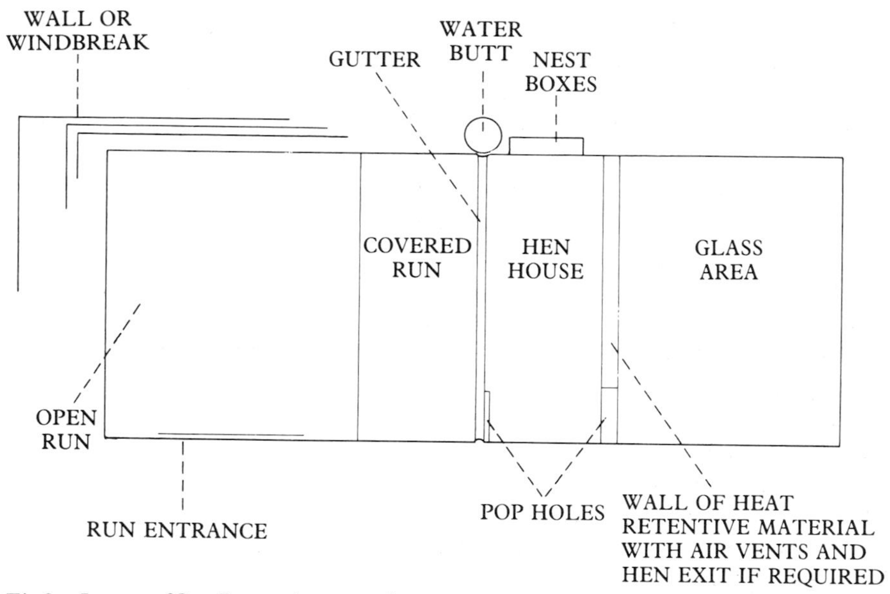

Fig 3 Layout of Sun Run, using as much or as little camouflage as required.

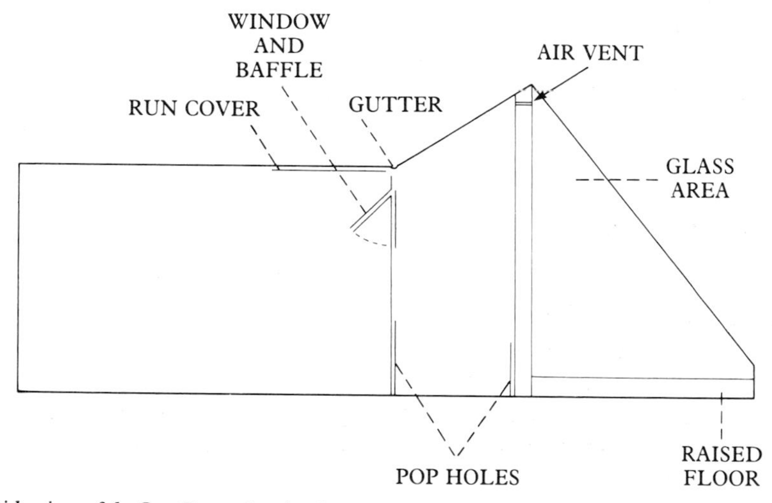

Fig 4 Side view of the Sun Run, showing how a slanted roof can be used to good effect for water collection.

in an open run it has been mentioned already that the larger the area the less height needed to restrict the birds, but in a covered run a good 3– 4ft (90–120cm) should be allowed. Without going to extremes, the more confined the area the more important becomes the run height, to allow free movement for the birds. A door or opening at the opposite end to the house will give easy access to water or feed troughs.

Materials When using wooden sheets or planks, don't forget that nails can split the thinner, cheaper wood and are more difficult to remove if the need arises. Screws allow parts of the house and run to be taken apart, particularly for cleaning or to replace a damaged section.

The Sun Run

If the chickens are to be kept in a town or city garden, it is not difficult to camouflage the run by combining it with gardening activities, to their mutual benefit. The illustrations show how the siting of a glass area, which can be used to grow salad crops and tomatoes, can benefit from the accumulated heat of the sun and the night heat boost from the poultry. The hen house must be closed at night.

If lighting is required, a qualified electrician can provide a safe passage for outside wiring. The economics of that will depend on how large an area is available. The perimeter hedging can increase in height along the run area, either by clipping or planting different species of herbs. It then performs the additional benefit of a wind break and sunshade.

A rotation of run area is possible when there are no crops in the glass area. Sunflowers, for winter use in the chicken run, can be grown beside the house for extra effect.

4 Husbandry Systems

FREE RANGE

There is only one true system of free range, and that means exactly what it says. The birds have total freedom to wander where they will, eat whatever they find, and lay their eggs in the most inaccessible places possible! All other systems of poultry keeping restrict that freedom in one way or another, and to varying degrees. The management of completely free range birds needs little mention here. Its legal definition, however, is becoming increasingly difficult.

Committees reviewing the whole poultry industry are currently and constantly investigating welfare issues relating to all sections of the industry. However, due to the complicated legislative arrangements between the British and European governments there is as yet no definition of free range on the statute books. However, a guideline set down to govern the sale of free range eggs, and one which has been accepted by a court in a prosecution taken out by a Trading Standards Officer in Shropshire says that hens need only have '. . . regular daytime access to land . . . in a density of not more than 150 birds to the acre, and from which the hens obtain a significant proportion of their diet from the greenery, grubs . . .'.

Anyone selling free range eggs is advised to check with their local Trading Standards Office as to their interpretation of the as yet unclarified position of the law. A copy of the Trades Description Act is obtainable from local libraries. A clear, up to date assessment of the situation might be forthcoming on application to the office of the local Euro MP. The relevant EEC Regulation is No. 2722/75, and deals with the labelling of eggs with their method of production.

Anything less than free range, therefore, must be referred to as semi-range, range-run, or other system. All these systems, other than the traditional farmyard laissez-faire, will be a compromise between allowing birds outside access, and control over their lives. No backyarder, though, would think of keeping poultry in the manner of battery cages, the other end of the scale from free range.

Unlike commercial poultry farmers, whose systems vary but little on the whole, all small scale egg and meat producers adapt the management systems to meet their own needs and opportunities.

NOTE

Any two or more of the following systems can be alternated, according to the seasons, or combined, according to prevailing circumstances.

RANGE

The birds can be housed in a movable ark, but be unrestricted for most of the day. Although one of the most 'natural' forms of husbandry, it is very labour intensive and, although work is said to be cheaper than food, today's poultry keeper could as easily think that work equals time and time is money!

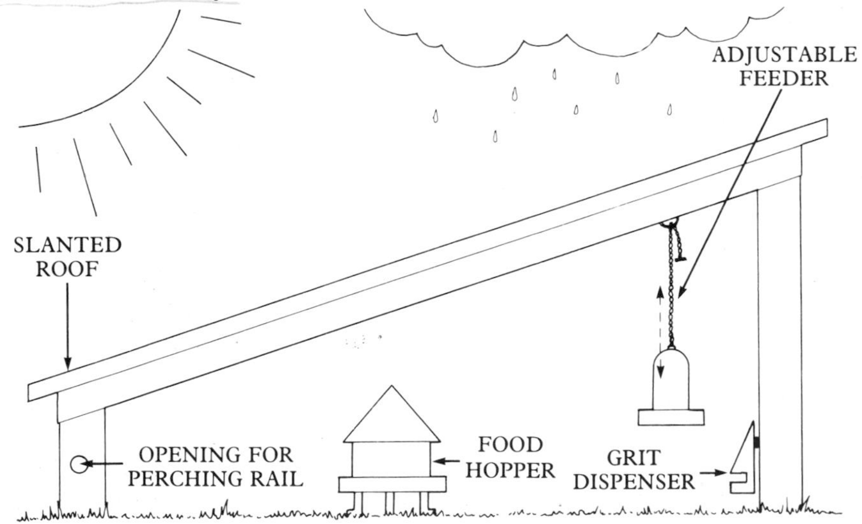

Fig 5 Shelter and shade without elaborate housing. Protection from predators can be given with the addition of netting nailed to the posts.

A range system must be away from a yard area, which makes regular inspections essential. Predators are a problem and birds should be shut up at night, sometimes a tedious chore last thing in the evening. It is often difficult to persuade the birds to use the official nest boxes, and by leaving them shut up too far into the morning, broken eggs and subsequent egg eating can occur. It is a fallacy to think that egg eating only happens with confined stock. It can just as easily start in range managed birds.

Those practising a range system must be prepared for inevitable loss of eggs and the occasional loss of stock. However, because of the birds' ability to forage for a large part of their diet, and the lower cost of fencing, netting and litter, it could be the ideal system where the land is available.

In an area suitable for range where no housing is available, a makeshift shelter can be constructed. Its use, though, is restricted by the lack of protection from predators. A simple wall of netting could be added for temporary range accommodation.

DEEP LITTER

The original concept of deep litter was imported from America in 1949. Birds were housed in controlled-environment buildings with the freedom to move about but with no access to the outside. The idea was that an initial starting litter of chopped straw, sawdust or peat moss was put down to about 3 inches (7.5cm) over a thin layer of horse manure. Although other manures could be used, horse was considered the best. When mixed with the bird droppings it activated, by bacterial action, a breakdown of the whole mixture into an ongoing build up of dark brown, dry material which could remain in the building for up to two years.

Suitable ventilation was required, to prevent the strong ammoniacal smell from becoming harmful to the birds. During the course of the first year the depth of litter could build up to 12 inches (30cm). The litter should never become wet or even damp, and were this to happen it would be an indication that all was not well with the system. Either the ventilation was not right, the building not properly insulated or the litter too shallow and the number of birds too few. An application of lime (not more than 1lb (.45kg) per 20 sq ft (1.86m^2)) was required every few weeks.

This system, in its pure form, is not much practised now, but a variation on it is a popular alternative to the original controlled environment housing.

The Ministry of Agriculture's Experimental Husbandry Farm at Gleadthorpe has been carrying out trials on straw yards and aviary husbandry systems which, if perfected, will aim to supercede the deep litter system and make it economically possible to reduce or replace the much criticised battery system.

The backyarder, however, can have his or her own straw yard system by combining the indoor housing, but with more frequent cleaning out, with daytime freedom for the hens, either in an enclosed scratching yard or a nearby piece of ground; a very good compromise between complete enclosure and total freedom, particularly in winter when a degree of control might be required regarding extended lighting. It is also useful when space for a permanent run is not available but some yard space is.

The birds can be housed until after mid-day, when in winter the earlier lighting will have extended the day, and they will have laid their eggs.

Most egg laying will be completed by mid to late morning. They can then be free to scratch and visit the dust bath before returning home at dusk (which they will do – provided the shed door is left open!). A regular feed of scattered corn at that time can encourage them to return to base.

This system works best for layers, but not for young stock or pullets just coming into lay. The pullets should be allowed to get into the routine of laying first, and have time to regard their accommodation as 'home'. There is little point in allowing fatteners out. They will not benefit greatly and the practice is guaranteed to reduce the time in which they will attain killing weight. Instead, greenstuffs can be provided in the house and a shovelful of earth given daily.

The stocking rate in a semi-deep litter system can vary according to the amount of time they are on range. A guide of 3 sq ft (.28m^2) per bird allows a slightly higher rate than if they were to be confined permanently. During winter, special attention must be given to providing greenstuffs to distract them and reduce the possibility of an outbreak of vice.

STRAW YARDS

This system can be used either with an access from a building, as described in Chapter 3, or merely with a netted yard and a covered end. Straw or other available litter is put onto the earth and the birds have access to it for all or most of the day.

There is one obvious problem with a totally uncovered yard, with no sheltered area available, and that is the weather. Wet, mucky straw is not conducive to good health. Apart from encouraging disease, it will often stick unpleasantly to the underside of the birds' feet in wet or damp weather. It will also result in dirty eggs. Ideally, therefore, the strawed areas should be covered or the straw raked up and removed regularly. A cover can be as rough or as sophisticated as circumstances allow.

Straw yards cater for the almost insatiable need that chickens have for scratching at the earth and, as with a semi-deep litter system, can be used successfully with growing stock or pullets as well as layers. With slightly more protection from the wind, rain and sun, the area could be made suitable for layers with no other housing, provided that perches are possible. Birds are more likely to suffer from stale air in an enclosed house which their owner has made 'cosy' than from exposure to the severest of winters.

If the area is the one and only housing for stock, the chief problem will soon be one of a fowl-sick run. It will become more prone to harbouring disease and worm infestation. It is therefore advisable to rest it annually and remove the accumulations of dirty litter and droppings from beneath the perching area. A few handfuls of lime raked over the ground helps to break down the remaining droppings.

Earth runs can be dug over before the next batch of pullets arrive. A quick growing crop of mustard is said to 'disinfect' soil if dug in when still green. Soot and salt can also be scattered in the run and dug in to help keep the area 'sweet'.

GRASS FOLDS

Another rather labour intensive system for anyone keeping a few hens in the back garden is the fold system. Although it can be operated without an enclosed run, it is more usual to have some means of restricting the scratching activities of the birds, and rather undermines the point of a fold system if grass is not allowed to rest periodically.

If commercial advice about the necessary grass seed is heeded, this can be very expensive. Even on very good grass folds, layers will need additional feeding. The advantages of such a system, however, where a house and run are moved frequently, are obvious.

Fresh ground will minimise disease associated with chicken-sick land, and the birds will not have enough time to do permanent damage to grass. Such damage can be reduced by having the run floor consist of large gauge netting.

Better use can be made of space if the folds are moved endways as opposed to sideways, thus allowing the roosting quarters to occupy the last grazing area.

If on range, frequent moving of the fold house discourages birds from laying away. They will have less time to look round and find a spot in which to lay their eggs. The only thing that will stop a determined hen from laying away is to prevent her from getting to the hedge, or wherever she has taken up residence. It is better not to give her the opportunity to find the spot in the first place.

Folding is a useful system for preparing ground for a garden. If the runs are left in the same place for long enough, the birds will reduce the area to bare ground which can then be forked over behind them. Again, some lime sprinkled on the area after it has been used will help to rot down the droppings. Liming is also useful where the area suitable for folding is limited and the birds have to return to the same piece of ground in a relatively short time.

PERMANENT ALTERNATE RUNS

If operated with care, and the room is available, a central housing area with two or more runs to be used alternatively, is a most satisfactory way of keeping poultry. By not overgrazing, a reasonable amount of grass can be maintained with less opportunity for weeds to build up in the bare patches. The different runs can also be used for stock at different stages of growth. A good rotation will ensure less build up of parasites and provide excellent rest periods for each run. Even the

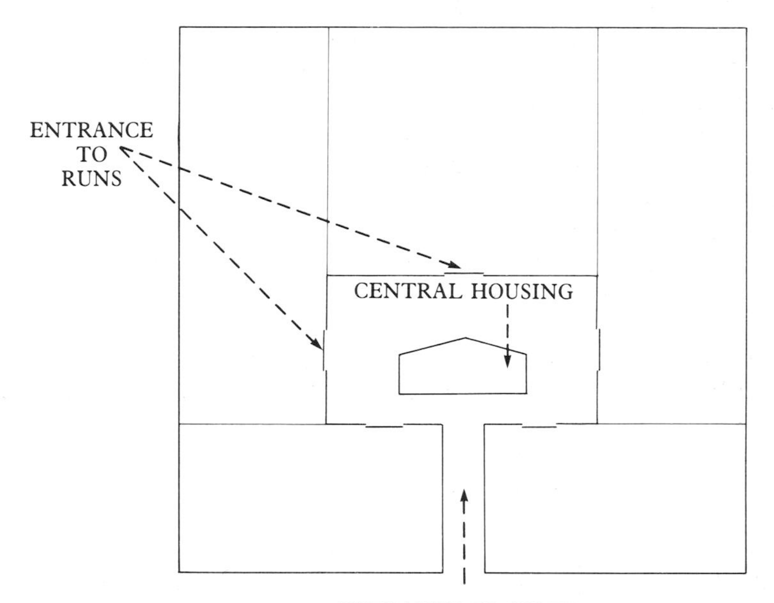

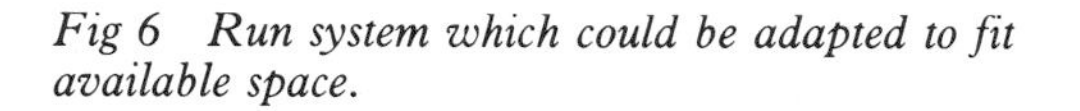
Fig 6 Run system which could be adapted to fit available space.

housing can be rested occasionally when a makeshift shelter can be provided in one of the runs, either during the moult period or at the start of a new season for young pullets not ready to begin laying.

The initial disadvantage of such a system is the expense of providing what could be large areas of escape proof fencing. However, this type of system, if done properly in the first place, is likely to last for many years and in the long term will save hours of moving houses about, worrying about the risks of disease build up, and ease management considerably.

The more convenient the system, the more likely it is to work to the benefit of all concerned. Apart from the usual tasks of egg collection, feeding and general husbandry practices, it is a worry-free system.

5 Management

The advantages of good management are twofold. In the first place, it usually makes stock keeping easier, and therefore more enjoyable. In the second, it will no doubt contribute to making the economics easier to assess, thereby showing up any profits or losses.

Good management consists of ensuring that input, i.e. initial capital outlay, replacement equipment and food, is outweighed (or at least equalled) by output, i.e. eggs, meat or both. Except for the more complicated and slender profits to be made financially from selling ornamental or showing stock, most people will want their enterprise to show some returns for their trouble; over and above the joy of keeping poultry that is. The necessity for a profit situation to happen is often a pity. There is no limit to the amount of expenditure which could be made on housing, equipment, and so on. Even if the enterprise is a hobby, or an exercise in giving the children 'something to do', there must be some satisfaction in getting the best from the stock and producing eggs which, if not cheaper than bought ones, are certainly to be considered of better quality.

RECORDS

Keeping records sounds a tedious sort of occupation, but working on the principle that simple is best, there is no need for them to consist of anything more than an ordinary notebook or desk diary. There will always be the time when it is useful to look back over what happened to the stock during the year, and no matter how well you think incidents are indelibly imprinted on the memory, they rarely are.

The kind of information which is useful to have at a later date includes the date and price of all stock and equipment; the cost of feed, and how long it lasts (which can be compared later to the number of hens or broilers eating it), and separate records for each group of poultry. Stock deaths should be noted, as should veterinary expenses, the number of eggs laid and the dates on which the totals were taken. Provided all the different sets of information are listed separately, the accounting can be left for another day.

Egg Totals

An easy way of keeping daily egg records is to hang a board and chalk in a convenient place and mark down the daily total. The number of laying hens should be recorded at the same time and any deaths or disasters noted. Then, either at weekly or monthly intervals, a more permanent note can be made in a diary and the board rubbed clean for the next period.

At the same time it is interesting, if not strictly necessary, to record comments about, say, the amount of greenstuff available, the amount of rain and its effects, any repairs that had to be carried out, or other events. So long as a basic record of actual eggs laid is made, with the dates when food was bought, the mathematics can wait for some dreary December evening when there's nothing much on the television!

Consistent laying records are not always necessary throughout the year, though it is useful to know when the first egg was laid. Perhaps recording one week in four during the peak laying time would be enough to indicate that all is well, while a closer watch can be kept towards the end of the season when the lowering of production will indicate the start of the moult.

Breeding Records

If replacement stock is to be home bred, it is essential to mark parent birds. Leg rings are the simplest, least complicated and cheapest. They can also be removed for showing. There is an excellent selection of colours and sizes of rings available from suppliers, and by keeping a note of the colours, and whether the right or left leg is ringed, much time and frustration can be saved later on when it comes to culling or assessing performance. Some breeders have a particular colour for a specified half or quarter year period. This will be particularly important when breeding for show purposes. It is sometimes easy to identify birds on home territory, but if more than one is being taken to a show it helps positive identification when allocating numbers or instructing an assistant. The modern plastic rings are very easy to put on and remove and cause no aggravation to the bird whatsoever.

If a particular strain of bird is being singled out, more careful records need to be kept of their performance. This usually means eggs laid, the age when the pullet first came into lay, and when she finished. It is not worth anyone's trouble to breed for meat birds, when it is done so efficiently and effectively by the commercial companies. It is arguable as to whether it is worth all the trouble of breeding replacement pullets, given the relatively low cost of buying in purpose-bred stock, except for showing. For this reason, it is well worth proving by means of a few basic sums whether this is the case or not.

Lighting Costs

To evaluate the advantages of using extended lighting, there can be no alternative to jotting down all initial outlays on time switches, bulbs and fittings, and assessments of electricity usage from the regional Board. Egg production, though not at its peak, should justify these costs during the winter months. The only way to find out is to add up the costs over a given period, including food, cost of stock and replacement equipment, and divide the number of eggs into the total. Many a nasty shock is realised by poultry keepers when the real cost of their eggs is unfortunately revealed.

USE OF RESOURCES

Apart from records, good management means getting the best from what is available to the individual poultry keeper. For example, it would be senseless to pay £50 for a hen house and then fill it with scrub Bantams. Conversely, it does not make sense to expect highly bred stock to perform well under bad housing conditions. It is also worth buying the best equipment that the budget allows. The ever increasing number of suppliers make it easy for the newcomer and old hand alike in the variety and range of drinkers, feeders, houses, netting, incubators, etc. Every so often, a new range is introduced which is inevitably designed to make the job easier and more efficient.

Certain management practices, or refinements on basic husbandry skills, can further contribute to minimising that imbalance of input against output. Not all practices associated with commercial production are necessarily bad or inhumane. Do not be put off by such words as 'controlled environment' which are fraught with visions of battery houses. It is perfectly possible to rear large numbers of birds for productive ends by humane methods. Controlled environment merely means that the immediate surroundings are controlled by someone other than the birds themselves. Should the birds control their own environment, humans would no doubt be very short of both hens and eggs.

ENFORCED MOULT

The natural moulting time for chickens is from midsummer to early autumn and is merely the replacement, by degrees, of new feathers for old.

Normally, the birds stop laying shortly before the moult actually commences, though some will continue to lay after it has started and begin again before it finishes. The duration of the natural moult is variable, but on average it takes between 8 and 12 weeks, and always in the same feather order.

There is such a thing as a partial neck moult, often in birds hatched during October to February, and it can happen before the pullets even start to lay. Birds hatched in March or April are more likely to lay for about 12 months or so before they moult, having begun laying in September. Usually, the earlier the birds go into a natural moult, the more likely they are to be poor layers, and should be marked accordingly and watched during the ensuing season.

The idea of an enforced moult therefore, is to arrange, by management, the exact time when it will take place, thereby giving the birds their rest and bringing them back into lay at an 'unnatural' time. Another reason for forcing a moult is for show management, or to allow for the laying and hatching of breeding eggs. Also, it is often convenient to allow egglessness to coincide with a family holiday or busy period.

Enforced moults were practised widely by farmers in the days when large flocks of range birds were kept on a commercial scale. The birds

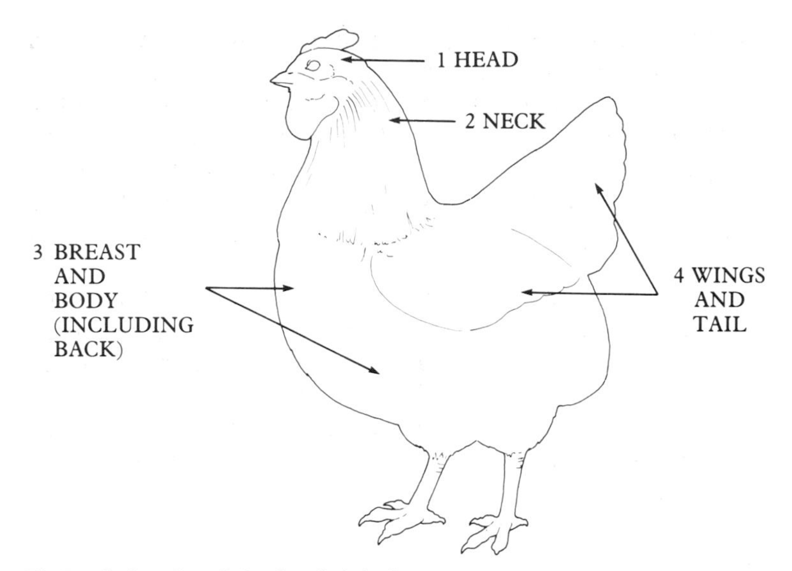

Fig 7 Order of moult for the whole body.

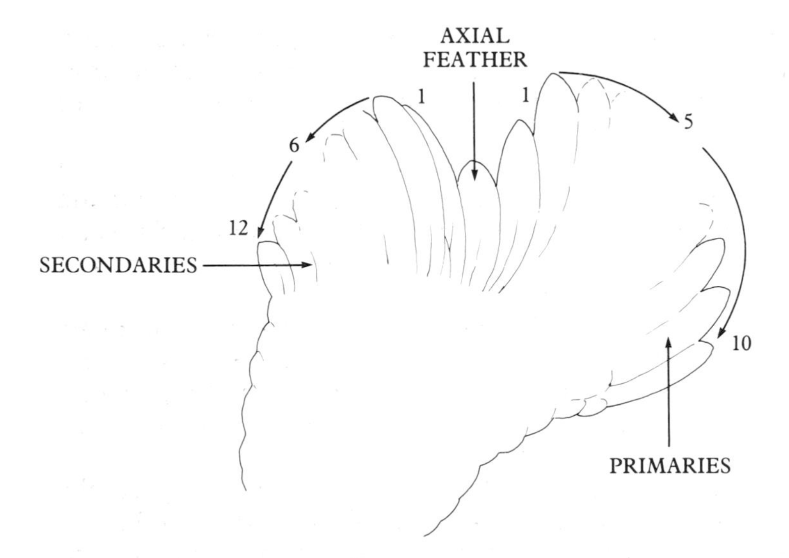

Fig 8 Wing order of moult. Secondaries do not necessarily follow in order 1 to 12.

would be brought in from the fields, confined in a house and fed only a small handful of oats twice a day plus some greenfood at midday. Epsom salts were added to the water, which was fresh and readily available. Once the new feathers were visible as stubby quills, the hens were given high protein diets with cod liver oil added, and brought quickly back into lay. The key to success was to make sure that the moult was complete, and not partial, or the birds would lay a few eggs and go back into moult.

A similar practice can be carried out with deep litter or straw yard stock. By August some birds will already show signs of moulting, so by restricted feeding for up to 12 weeks, they can be brought back into lay by October, and go straight onto artificial lighting.

Once the moult is over, a high protein feed is essential, and the extra additive of poultry spice, available from most of the larger equipment suppliers, can ensure a supply of vitamins and minerals.

LIGHTING

Left to themselves, hens would only lay during spring and summer, when daylight hours are lengthening. This might be fine in an all-range situation, but it doesn't help with the economics involved for those hoping for year-round eggs.

Provided adequate food is available, and the birds have been through a natural or enforced moult, there is no reason why first, second and sometimes third year birds cannot be brought back into lay during early winter, with the help of artificial lighting. For the novice, it is easier to manage pullets than year-old hens.

Pullets should be at least 20 weeks old, to have allowed adequate body growth before the demands of egg laying begin. Second year hens should have gone through a full moult during August and September, or overlapping into October. The lighting plan can then begin during October.

The hours of natural light during each 24 hour period throughout the year varies, both due to the seasons and prevailing weather conditions. The shortest days are in midwinter and, under natural conditions, the increasing amounts of daylight during the spring stimulate the reproductive organs of the birds. Messages are passed from the brain to the

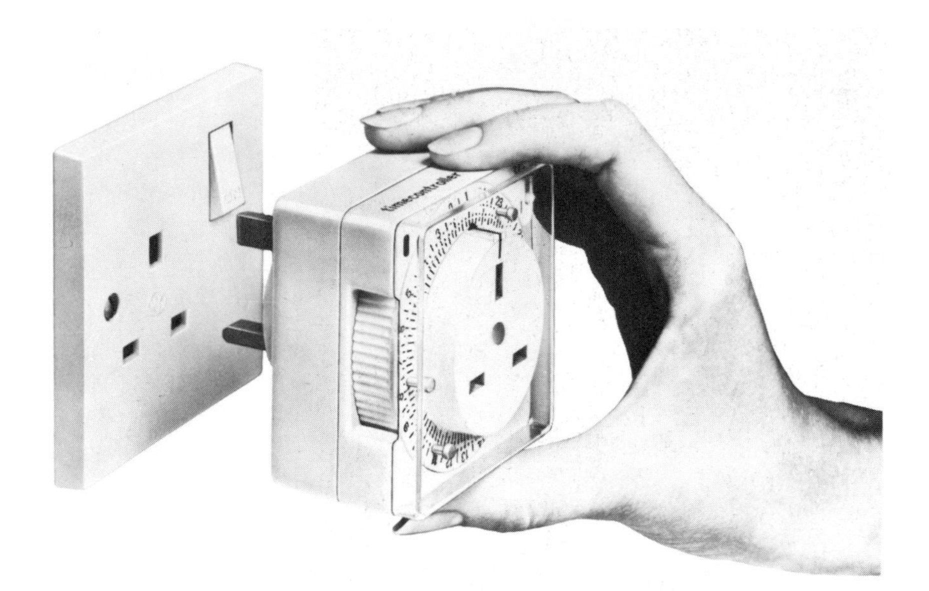

Time controllers in 24-hour and seven day versions are available to regulate a winter lighting pattern, and can be used in standard square pin sockets.

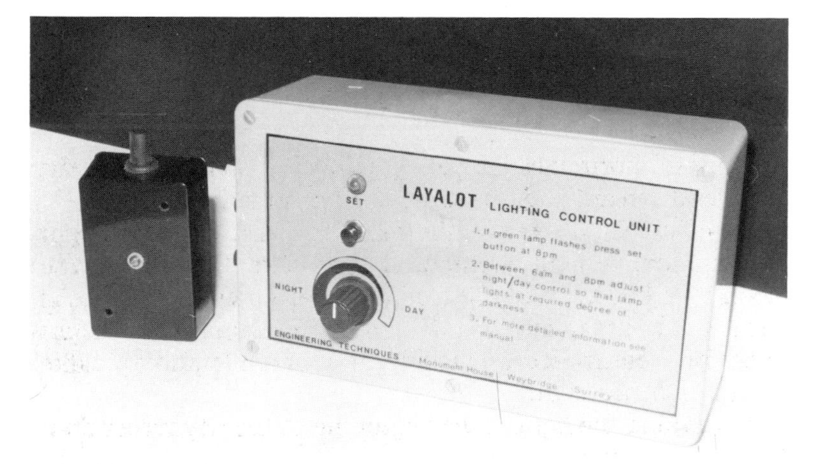

Some lighting controllers incorporate a sensor which can measure the light levels and compensate if necessary.

Ideal environment for extended winter lighting.

pituitary gland, and hormones are produced which then travel to the ovary in the blood stream and stimulate yolk production.

Stock then are likely to respond to the stimulation provided by a light bulb, which can complement the natural light, either to extend the length of daylight or boost the strength of light on a dark winter's day. If natural light is weak, it can lose its effectiveness as a stimulant. Light intensity therefore, controlled artificially, is a variable and valuable component in a lighting plan.

It is better to extend the light at the beginning of the day rather than at the end, as it has a more consistent effect. If no natural dusk light penetrates the housing, a dimmer switch incorporated into the system would allow the birds some warning that it is bedtime, as a sudden withdrawal of light can cause the birds to crowd and possibly suffocate. Convenient and reasonably priced time controllers are on the market and are available in 24 hour and 7 day versions. They can be plugged into any normal 13 amp socket. Another, more expensive but very effective aid is a controller, which has a sensor unit fitted and will automatically put the lighting on if the ordinary daylight is too low, perhaps during cloudy or foggy weather.

Beginning with 8 hours of light per day, this can be increased by 6 or 7 minutes a day until a steady 18 hours of light per day is reached. Some commercial producers begin with 18 hours, then drastically decrease to 6 hours per day for between 2 to 18 weeks, and then begin the weekly climb of 45 minutes to bring the total back eventually to 18 hours. Obviously, the second scheme would need to begin earlier in the season before the first, and is in effect an enforced moult.

CULLING

The importance of not carrying hungry passengers cannot be stressed too strongly. Some birds seem to delight in devouring huge amounts of food and giving very little in the way of return. These birds are luxuries that the small scale producer and commercial farmer can well do without.

Culling can be one of the less pleasant aspects of poultry keeping, and often the most difficult. First, an assessment of hen's worth must be made, and then the remedy or method of disposal decided upon.

Laying fowls are culled for several reasons. The main one is that they are no longer laying sufficient eggs to justify their keep, either because they are too old, or because one or two birds in the flock are unthrifty and bringing down the egg average. These are the birds which must be sought and disposed of. The age problem is often one fraught with sadness. The smaller the number, the more likely the culling process is to be emotional. So often, when only five or six birds are kept, they quickly develop characters.

Culling is best done at the time most likely to show up the poor layers. This can either be done during the laying season, by means of trap nesting, or during the moult when it is likely that birds which moult very early need culling. During the moult it is possible also to see whether a bird is still very fleshy, which means her food has been going into her body and not into egg production. Good layers rarely put on much weight during the laying season.

Trap Nests

A trap nest is simply a method of devising a way for a bird to enter the nest box without hindrance, while preventing her exit when she has laid her egg. In order to prevent the start of an egg eating habit this is a process which has to be strictly supervised. When released, the bird should be marked in some way, either by leg ringing or marking her with a waterproof marker, and a record kept of the amount of times she visits the nest and the eggs laid. This is also useful information to have on hand if it is decided to begin breeding. Any bird who lays comparatively few or no eggs should be considered as dispensable. A further check on condition can be made by removing the bird at night, when at perch and in a drowsy state. Sometimes a hook can minimise the disturbance caused by getting too close to the birds. The crop should be full, assuming that it has been possible for the bird to eat during the evening. Any with empty or half-filled crops should be checked again with the trap nest records and definitely culled.

The same secondary test can be carried out with early moulters. Any bird, for example, moulting in June is likely to be out of production for the rest of the season. It might be necessary to cull for persistent broodiness if it is not required. Disposal of these types of birds, though, is usually no problem as other breeders might be willing to pay for a good broody.

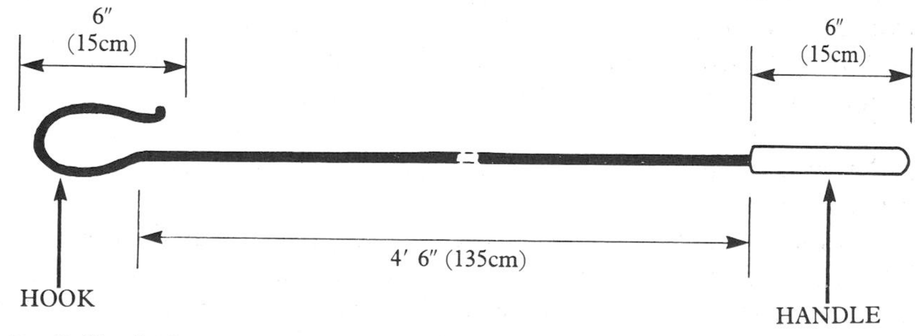

Fig 9 Culling hook.

Trap nests must not be too complicated. Pieces of string or wire getting in the hen's way might put her off, and any loud noises or dangerous guillotine effects will not encourage any bird to use them more than once.

If the birds destined for disposal are to be eaten, and this aspect could well have influenced the breed of chicken kept, care must be taken to kill them humanely. Not everyone can kill a chicken, either by wringing its neck or by any other means, so it might be necessary to find someone who can. Poultry waiting to be killed should be watered but not fed during the day before slaughter. If chicken rearing for meat is part of the enterprise, it might be worth investigating the cost of a humane poultry killer, which decapitates the bird with a guillotine-type blade. There is rather more blood in evidence with the second method, though it is better to do the job properly first time than cause the bird distress and pain by failing to dislocate the neck properly.

When killing, prepare for the event in advance. The birds need to be plucked as soon after killing as possible, as the feathers come out better while the body is still warm. Feathers and blood don't cling to waterproofs as badly as jumpers and other clothes, and can easily be hosed off if necessary. Disposal of the feathers, heads and feet should be by burning or burying. They should not be left lying around as an encouragement to vermin and insects.

OBSERVATION

A final management note concerns the value of observation. A few minutes every so often spent watching the stock and observing their normal air of contentment will help to develop the sixth sense to know when something is wrong. The birds themselves are the best barometers of what is best for them.

6 Feeding

No chicken will lay well if it is not fed adequately. It will survive and look healthy on a meagre diet, but all hens, pullets and fatteners need plenty of protein to attain full body growth and produce eggs and meat. Carbohydrates are required for energy, and vitamins and minerals to maintain health and combat disease.

FORMS

Food for the home producer comes in a variety of forms and a multitude of combinations, and very much depends on the functions of the birds. Poultry at different stages of their lives have varying nutritional requirements, and careful feeding is crucial to both performance and the slender tightrope between profit and loss. Basically, the backyarder is looking for the cheapest food which will give the best results. Whether it is cheaper and better to try to produce the protein needs at home, or to buy in ready-formulated feed is a decision not always easily made right at the beginning of the venture into poultry keeping. A useful compromise at first is to buy in pellets or whole grain and complement that diet by seasonal garden growth. The amount of work and time involved in producing sufficient feed for even a small flock of egg layers is not immediately obvious when first embarked upon, and it is not surprising that most poultry keepers buy either pellets, meal or whole grains.

Scraps

The idea of a chicken bucket under the sink is not a particularly good one. There is bound to be a high proportion of unsuitable feed, e.g. potato peelings, which only lie around in the run looking unsightly and being ignored by the hens. They are not at all keen on other refuse, such as orange or banana skins, or pea and bean pods. Often some meat or meat product pieces may be thrown in by mistake, which is illegal.

If household scraps in any form are to be used, they must be boiled for at least an hour, and any hotel or restaurant must have a Ministry of Agriculture licence if it intends passing on food waste to stock keepers. Though some salt is essential for health, it is not tolerated in any quantity by the birds, and there can sometimes be a high salt content in household waste. Large amounts of salt cause high water consumption and eventually death.

Any selected household scraps, i.e. vegetable waste, fish scraps, and so on, which are used can be fed with a balancer to dry the boiled mixture into a crumble. It is then easier to feed and not so wasteful, in that it is picked up better by the birds. Larger whole vegetables and greens which are to be hung in the run or building need not be boiled.

It is doubtful that the average household could maintain sufficient scraps to feed fattening or laying chickens. Even the potato, a traditional winter feed for fattening, has to be fed in such large quantities – 4oz (112g) of cooked potato for 1oz (28g) of mash – that it encourages over-feeding and is very time consuming for the owner to prepare. Laying birds must not be allowed to become fat, and a large part of what will probably be put out for the chickens should more properly be put on the compost heap.

Grain and Pellets

Seeds, grain and insects have formed the basis of the chicken's diet for a long time. Grain is fed either in a whole form – barley, wheat, oats or maize – or ground down and fed as loose mash, pellets or smaller 'crumbs' for chicks. Easily the most convenient are the pellets which are bought in half hundredweight (25kg) bags.

The amount of protein and other relevant information will be marked on the label, and advice about which feed to choose can be sought from the supplier, who will need to know the age and function of the birds. All commercial chick crumbs also contain a coccidiostat as an aid in the prevention of coccidiosis, and most fattening pellets will have a growth promoter added to them. If there is any additive in the feed, it will by law have to appear on the label attached to the bag, which will help anyone not wishing to buy such preparations. Warnings are also carried on the labels which indicate the necessity to withdraw medicated feed

BABY CHICK CRUMBS (0–6 weeks)	
OIL	2.25%
PROTEIN	18.50%
FIBRE	4.50%
ASH	6.25%
GROWERS MEAL (Growers/Rearers 8–16 weeks)	
OIL	2.00%
PROTEIN	14.00%
FIBRE	5.00%
ASH	5.75%
INTENSIVE LAYERS MEAL (Layers 18–20 weeks onwards)	
OIL	2.25%
PROTEIN	14.50%
FIBRE	5.00%
ASH	12.00%
POULTRY FATTENING (16 weeks onwards)	
OIL	2.75%
PROTEIN	15.00%
FIBRE	4.50%
ASH	5.25%

Example of commercial feed compounds. (Information supplied by Allen & Page Limited of Norwich.)

either seven (or other specified number) days before slaughter for human consumption.

AMOUNT

On average, six birds will need around 12lbs (5.5kg) layers' mash per week, allowing a little extra during the winter. Methods of feeding will vary, but if ad lib feeding is practised, it is worth investing in a vermin proof hopper, preferably one which the birds can learn to open themselves, so making it useful on range as well as inside. Rats, mice and birds are only too willing to take advantage of any food left lying around, and the hens themselves do not clear up dropped and stale food in the hopper area. The feeder should have a lip to prevent such spillage. When buying a feeder, it might as well be the most useful and efficient one possible, and there is certainly an excellent choice available.

If the protein level is low, more food will be consumed in order to get the required amount of nourishment, and the extra energy taken in will be stored as fat. A ration slightly low in protein may not impair initial growth but can restrict full body development and interfere with the egg laying process.

WHEN TO FEED

When and how to feed will depend very much on what system of housing the stock is kept under. If they are housed in a building or range house, and let out during the afternoon, the ideal is to allow free access to feed during the morning, closing it either at midday or when the birds are let out. They can then be given a grain feed in the afternoon to encourage them to return home and to allow them to fill their crops before dusk. No more than ½oz (14g) per bird of grain need be given each day as more will upset the balanced ration so carefully formulated by the feed company. Many people like the ritual of scattering the corn down for the hens, while others just think it is 'good for them'. The practice certainly has one advantage in that it gives time for observation. If any bird lags behind the others, or is off her food it is easier to spot them.

Another useful method of feeding mash and grain is to use a wet mash in the mornings – using warm water in winter – and the grain in the afternoon. This allows more control over the exact amount of food given than an ad lib feed. The same system can work as well the other way round – grain in the mornings and mash late afternoons. Either way, the ration of between 4 and 6oz (112–170g) per bird can be split between the two feeds. As there is no such thing as the average hen (compare a Brahma to a Japanese Bantam), the 'ten minute rule' is not a bad one to follow. This means that whatever amount of food the birds clean up in ten minutes, is the correct amount required. A much debated opinion, but one which seems to have survived years of criticism.

Owing to the perishable nature of mash and pellets, there is always a date stamped on the feed bag label, giving the last date when the vitamins and minerals will be at their best. This isn't quite so crucial to birds with access to other foods as to an intensive chicken, and obviously the values do not immediately decrease on a specified day. It could, however, be important in chick crumbs, and the dates should be checked when purchased.

STORAGE

Once bought, it is essential that feed is stored in a vermin proof and dry container. Commercial enterprises with a big turnover of feed will not worry about storing each bag in airtight containers, and are more likely anyway to receive their food in bulk. But if only a few pounds a week are being used out of a half hundredweight bag, there is no better way of keeping it safe than in a dustbin – a clean one, of course. To make sure that new food isn't tipped in on top of the remaining old, dustbin liners can be used so that if there are a few pounds left when the new bag arrives, it is easily removable. Whole grain does not deteriorate in the same way as compounded feed and therefore has a longer shelf life. It still needs protection from damp storage conditions.

Wheat

Wheat has always been a popular feed for hens, and with a protein content of around 8–10 per cent, it is valuable for laying hens. The bran, the outside covering of the grain, is not suitable on its own because of the high fibre content. When buying wheat, make sure that the kernels are plump and firm, not hard or shrivelled. There is little or no feed value in bran and not much more in a poor kernel which was probably dead when harvested. Protein content of grain is variable, and depends very much on the conditions under which it was grown. The different varieties bred and grown for their high yields can also have slightly different feed values. Methods of assessing protein can also vary.

Barley

Barley is widely fed to hens in parts of the country where it is freely grown, and has roughly 7–8 per cent crude protein. Hens do like it, but due to the high fibre content, it does not always agree with a hen's digestive system. It should not be fed exclusively.

Hens do very well if folded on stubble, that is free to forage on fields of both barley and wheat after the combine has been in. Care should be taken though that pullets who have only just begun to lay don't go into a partial moult if moved onto stubble range. Barley can cause second year birds to become fat and lazy if fed as a high proportion of the diet.

Oats

Oats are often used in a mixed corn ration, but they are very fibrous. They are fed commercially in meal form to birds being fattened for the table, and were traditionally fed to hens at the end of the moult. Sprouted oats are a useful feed but not all birds will eat them. Oats are usually crushed to aid digestion.

Fig 10 Scaffold for hanging greenstuffs and roots; a similar arrangement can be used for housed stock.

Maize

Maize is also used as part of a mixed corn ration, but is more often used in meal form. With a crude protein level of between 8 and 9 per cent, it has always been thought of as an indispensable part of the layer's diet, and an ideal feed for fatteners. It has also been associated with encouraging a deep orange yolk colour, the yellow pigmentation 'feeding through'. For this reason, it is not used by fanciers who show white plumaged breeds, though at other times of the year – particularly winter because of its high oil content – it can be used as part of the ration. Breeders of dark plumage breeds, or yellow legged varieties, can use maize all year round. For smaller breeds, the maize grain is usually kibbled, that is broken up into smaller pieces. It can also be flaked or made into meal. Maize has a high carbohydrate value, making it a useful winter feed. It can, however, cause scouring if fed in too large a quantity or on its own.

Greenstuffs

Almost any brassica crop can supplement the protein diet, and has the advantage of taking longer to eat than pellets or grain. This is particularly useful for housed stock, especially during winter. Cabbage type plants are always hung at about head level so that they don't get fouled and wasted. A permanent 'T' shaped scaffold with hooks can be left in a run or house for constant, year-round use. A similar system can be adopted if only loose leaves are available. These can be packed into a string bag and hung up. A ready supply of these can be provided from any garden, especially with the cut and come again varieties. The main cauliflower stem, if left in the ground after the flower is cut out, will regenerate and provide leaves for weeks. A few of the everlasting lettuce varieties are worth the garden space, always remembering that it will wilt quickly when picked and will not last as long as cabbage in a palatable condition.

Other garden surpluses, marrow (especially if left until the seeds develop), swedes, turnips and other root crops can also be impaled at head level. Root crops are completely wasted if thrown onto the run floor as they will rapidly become totally unpalatable to the chickens. When hanging up a large swede or parsnip, it often helps to cut it open or halve it to encourage the birds to peck out the soft middle and not be put off by the hard outer skin.

Weeds can be a useful addition to the diet, including chickweed, which is aptly named. Young chicks peck eagerly at it, but only so long as it is fresh. If given frequently it is said to be effective against coccidiosis in very young stock.

Comfrey is often grown for chopping and feeding to layers, though it is not easy to assess its exact feed value. In the past it has been used as an alternative feed to clover but it is very fibrous and should be chopped when fed rather than given as whole leaves.

It is never easy to assess feed values in greenstuffs, but their value lies in providing the extra vitamins and minerals required by in-wintered stock, and in 'giving the birds something to do'. The chlorophyl present in greenery ensures a dark yolk colour. It has been worked out that a laying hen would need to eat 22oz (.63kg) of cabbage per day to keep fit and lay her eggs, but she could only actually manage to eat 12oz (.34kg). It is as well that protein is available elsewhere!

Grass

If sowing grass seed specifically for grazing, advice should be sought from the seed company. The retailer selling lawn grass is not always able to help with information about the different varieties, but might be able to give the address of the wholesaler, who might be more knowledgeable. Timothy and meadow grasses are frequently recommended, and any preparation containing Italian ryegrass or cocksfoot is suitable. Some grasses seed early in the growing season. Fibre levels increase as the plant matures and old, stemmy grass will be ignored by the fowls.

Grass does not grow evenly throughout the year, nor is its feed value stationary. It is at its best nutritionally during April, May and June – assuming clement weather conditions. A very dry summer will deplete the autumn grass values. If grass is to form a large part of the diet, thought should be given to a means of fertilising the folds.

The previous year's fertilisation from the birds themselves will be of variable use and, unless densely stocked, not enough to ensure adequate spring growth. The birds should be kept off the grass in early February and March, otherwise grazing runs the risk of killing it off and weakening the growth for the rest of the season.

Relying on too much grass feed can result in crop binding, particularly with young stock who are not used to it. If no other feed is available when they are put out to grass it is almost bound to happen. They need to be weaned onto it gradually making sure that they have a morning feed of a familiar mash or grain.

Provided that lawn clippings are fed in moderation, they can also be profitably used as an additional feed. If the clippings come from a neighbour it is as well to enquire about any chemicals which have been used in lawn care during the year. Gardeners who don't keep chickens don't always think of these things. Again, too large a quantity fed at a time when no other food is available can also end in crop bound stock.

Cod Liver Oil

Not only a convenient source of vitamins A and D, but much used by the Fancy to ensure glossy plumage. Cod liver oil, the sunshine vitamin, is a winter substitute for the sun, but only a small amount is necessary.

A teaspoonful daily is adequate for a dozen small chicks, working up to an adult ration of 2oz (56g) for every 7–8 lbs (3–3.6kg) of mash.

Poultry Spice

Several recipes for this have been resurrected and are now sold by some of the larger equipment suppliers. Usually, if the diet is well thought out and varied, the extra vitamins and minerals are not strictly necessary but more of an insurance towards peace of mind. It is, however, used by the Fancy to bring show birds up to their peak condition.

Miscellaneous

The by-products of smallholding, in particular dairying, can be used to supplement the diet of poultry at all stages of their lives. Spare milk can be mixed into mash, but always fed freshly prepared. If the milk goes sour after it is mixed with mash the birds won't eat it, which will cause waste and disrupt their food intake. They will, however, eat it once it has soured on its own and formed into solid lumps.

A little milk in the feed of young chicks is useful, but due to the enormous amount of potentially harmful bacteria it can cause scouring and should be handled with nothing less than scrupulous care. Some fanciers feed bread or biscuits with milk shortly before a show to encourage extra plumage gloss. It is also used after a pre-show bath, with an extra spoonful of sugar.

If milk is used in troughs, they should be rinsed out before particles stick to the sides. There is nothing more calculated to put a hen off her feed than a smelly food container – apart from which it will increase the risk of disease.

Acorns, if gathered in sufficient quantities, are another possibility for a diet complement. They should be fed dried and crushed after first being shelled and boiled. Acorns should not be fed to very young stock, and only sparingly to layers. They can cause digestive problems and discolour the yolks, turning them from yellow to a muddy green.

Sunflower seeds can also be home grown, stored and fed as a winter titbit, as they contain good protein. Their fat content is seven times that of oats. Seed heads should be gathered just before the seeds are ripe and dried for a month or so before feeding. They can be hung up in a shed and the birds allowed to peck out the seeds.

Brewers' grains, a by-product of brewing, can be fed if obtainable. It can be mixed with other poultry rations and fed wet or dry. If purchased wet it must be used quickly or dried as it can sour and go mouldy.

Things to be avoided include sugar beet tops, potato shoots and rhubarb, as they all contain substances poisonous to birds. Most garden weeds do no harm, with the notable exception of the herb henbane.

If fish meal is obtainable, this again is a valuable protein food, but now very expensive. A small amount does not seem to taint the egg in any way. Tainting is often caused when droppings and eggs come into contact in the vent area, due to the very porous quality of the eggshell, and is rarely caused during the formation of the egg within the bird's oviduct. Fish scraps from the local fishmonger are a most valuable food and when fed (cooked) they do meet with wholehearted approval.

Grit

The uses of the two types of grit, soluble and insoluble, are frequently confused. The first is required to satisfy the enormous demand by the hen for sufficient shell-forming material, i.e. calcium, and also for strong bones and good growth. Between 8 and 11 per cent of the whole egg is accounted for by the shell, which is 97 per cent calcium carbonate. The usual cause of rickets in young chicks is the imbalance of calcium/phosphorus in the diet and insufficient vitamin D.

The second, insoluble grit, is to assist with the grinding down of food in the gizzard. The muscular action of the gizzard walls thoroughly breaks down the food by crushing it against the insoluble grit which stays there and is replaced as necessary. Not only is the digestive system severely upset by food which has not been through this process, but portions of food passing undigested through the body are obviously wasted.

A large bird will need something in the region of 4 –5 lbs (1.8–2.3kg) of soluble grit and 2–3lbs (0.9–1.36kg) insoluble grit per year.

Grit with a lime content usually comes in the form of oyster or cockle shell, or limestone. It takes about 48 hours for it to be broken down in the gizzard and then it requires replacing. Without adequate supplies of calcium a laying bird will draw on its own reserves, that is the stocks of

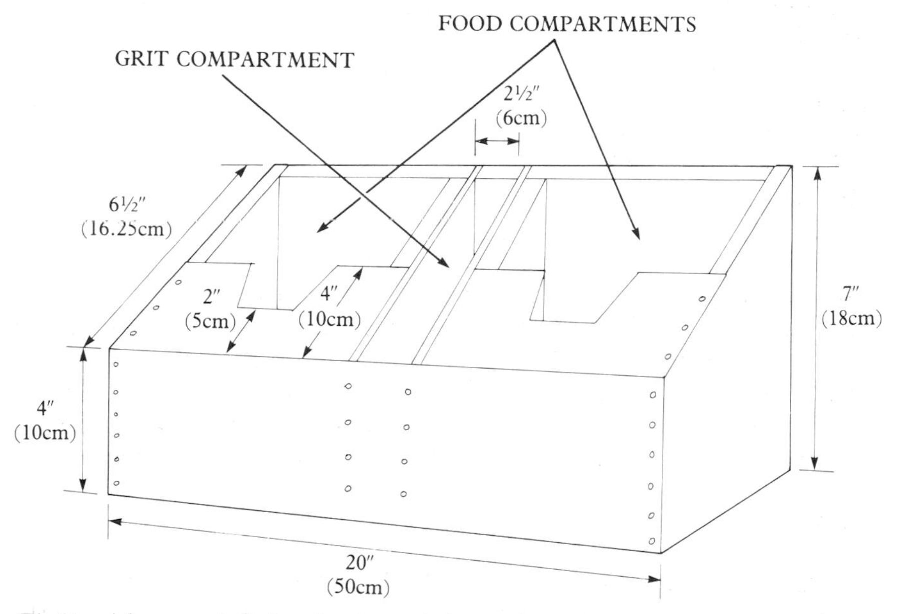

Fig 11 A home-made feeder, showing central grit compartment.

lime in the bones of the body, with the result that these become adversely affected. The grit should be freely available and checked regularly to make sure that it doesn't just consist of dust, which the birds won't bother with.

Too much calcium is as bad as too little, causing digestive problems due to the resulting high alkalinity. Often, if non soluble grit is unavailable, the birds will take up too much calcium grit in an effort to supply the required flints.

Where the birds are on range, or have regular access to the ground, the necessary flint grit intake is usually adequate. However, there is nothing to lose by making some extra flint available, especially in winter, to make sure that all is well.

A mixture of the two grits can be kept available to the hens in many types of special containers, in the ratio of one part insoluble to four parts soluble. If constructing a feeder for just a few hens, a grit compartment can be incorporated into the design.

Water

The provision of fresh water, available at all times during the day and night, is a much neglected one.

Since an egg has a water content of 70 per cent (with 48 per cent of it in the yolk), and the hen's body comprises some 60 per cent water, it is obvious that a bucket which has stood in a corner for a month and a half filled with a stale, murky liquid just won't do.

One of the new types of water fountain, easily plumbed in and made of cleanable plastic, is ideal where possible. The smaller the actual water area, the less chance there is of it getting fouled or dirtied during the day. If the birds have to perch on the edge of an open bowl, or it is in the line of fire during a scratching session, it becomes a chore to keep it constantly clean and palatable. It can also get tipped over and leave the stock without water for long periods. On range, the newer lightweight water containers take little time to keep filled and clean.

During the winter it is worth lagging any exterior pipes, to make sure that supplies aren't frozen for long periods. A kettle full of boiling water will raise the temperature a degree or two, though chickens aren't particularly fond of warm water. During summer the water should be placed in the shade where possible.

7 Breeding

The reasons for home breeding are twofold; to provide replacement stock, and to rear specialist lines for showing or sale. Additions to this can include hatching and rearing for purposes of education, or simply for the interest and joy of it.

The reasons for embarking on a breeding programme will influence how it is done. For example, if hatching for education purposes, an incubator with a good view of the egg is required, while for the smallholder or fancier it is more likely to be the broody hen.

METHODS

Before deciding upon the method, it is as well to consider the necessity of replacing layers or fatteners by home and often indifferent breeding. For the person who keeps a few hens for eggs, the cost, trouble and time expended on the project just do not justify the ends. Questions must be raised about the husbandry methods of modern poultry production; the results of specialist breeding by the commercial companies can as well be adapted to backyard methods as to battery houses. They can do the job of breeding more professionally and far cheaper. A traditional farmyard breeding programme has been one of allowing miscellaneous hens to produce offspring sired by miscellaneous cockerels (once referred to picturesquely as 'the common dunghill cock'). The females then lay for two months of the year and eat for twelve, and the males just eat for twelve.

In the true free range, unsupervised style, this is no doubt perfectly acceptable, but it is as easy to be economical and efficient as not. The cost of buying in scientifically bred stock to produce eggs or meat is still relatively cheap. The cost of a POL pullet has been almost unchanged for twenty years, while the potential usefulness of today's progeny is a credit to selective breeding.

Pullet Replacement

The most likely need for pullet replacement is when the breed is an unusual or rare one. Obviously the possibilities of breeding from hybrids are limited, to say the least.

Assuming that the female stock has been proved by means of home records, it is not worth considering anything less than a probably expensive cockerel, who is also proven or at least comes from good stock.

If a breeding trio has been purchased, and it comes from a reputable source, the burden of selection has been on other shoulders. It is a relatively simple matter then to get on with egg collection, hatching and rearing.

The basic rules of in-breeding, heredity and the quality and standard of stock which relate to other forms of livestock also apply to poultry. They are too varied and scientific to be mentioned here, and there are several in-depth publications dealing with the various aspects of genetics. However, it is worth mentioning one breeding factor which is peculiar to poultry; that of sex linkage.

Sex Linkage

One of the biggest problems facing commercial chick producers when hatching in large numbers first began, was the necessity of rearing both male and female chicks identically. That was, until it was possible to tell the two apart at hatching – a problem which is faced by laymen today.

There was no market for expensively reared males to the egg producers, and the males were preferred by capon and broiler farmers. Manual, and later machine sexing, was practised but sexing still presented a problem to the middle sized hatchery or one-man operation.

In 1919, Professor R.C. Punnett observed that the down of male and female chicks was consistently differently coloured in the chicks of certain crosses. An outstanding example of a 'gold' male and 'silver' female cross is the Rhode Island Red and Light Sussex first cross. The Rhode males transmit their dark colour to the female chicks and the Light Sussex females pass on the whitish or yellow colouring to the male chicks. Other gold and silver breeds can be used to produce similar offspring, plus a more complicated phenomenon based on the inheritance of the barring factor. For example, if a Black Minorca male is mated with a Barred Rock female, the male chicks will be barred, and the female black and unbarred.

The experiments were taken one step further in 1929 when a Cambridge University Professor discovered the auto-sexing principle, by which male and female chicks of the same breed could be identified.

The first of these was the *Cambar*, which was the Gold Campine crossed with the Barred Rock. Standards have since been accepted by The Poultry Club for these breeds and include the Brockbar, Legbar and Welbar.

Sexing methods a few years ago were predominantly those of manual vent inspection (a process developed by the Japanese and one which takes several years to perfect), but now many more of the newer hybrids are sexed by down colour, for example the Hubbard Golden Comet.

In hatcheries dealing with only layers, the male chicks are discarded and gassed soon after hatching. Feather sexing is another alternative and is practised exclusively by Ross Poultry in their broiler hatcheries. Broilers are of white plumage and therefore not suitable for down sexing.

Breeding for Show

The best way of breeding potential winners is by selecting parents from known winners or related quality stock. There is a lot of misconception about in-breeding, that is breeding closely related stock, which leads to the degeneration of that stock. In fact, if approached scientifically and knowledgeably, it is possible to keep the good points to the fore and the bad in the background.

Nearly all Bantams which are close to the Standards set down by The Poultry Club are near relations. The possibility of 'throwbacks' is minimised by only continuing to breed from the best, and never from second best. In this way, with the same objectives of breeding always in view with successive generations of a particular breed, a 'strain' can be developed. Certain characteristics within the breed will be emphasised and the stock which do not measure up to the high quality required will be culled or disposed of to a less discriminating breeder.

The ability to cull ruthlessly is often the key to success in showing. The same principle must also be applied if breeding for egg laying properties, after carefully assessing the performance of the parent stock. There are very few fluke successes in either appearance or performance. The top and therefore the most desirable birds in both fields are not obtained by luck or accident.

Education

Following the hatching process is now recognised as an ideal way of showing students the life process in the most practical way possible. There are several small, inexpensive incubators on the market which have clear observation tops giving full vision through the complete hatch, making them suitable for use in colleges and schools.

If a hatch is being carried out at an agricultural college there will no

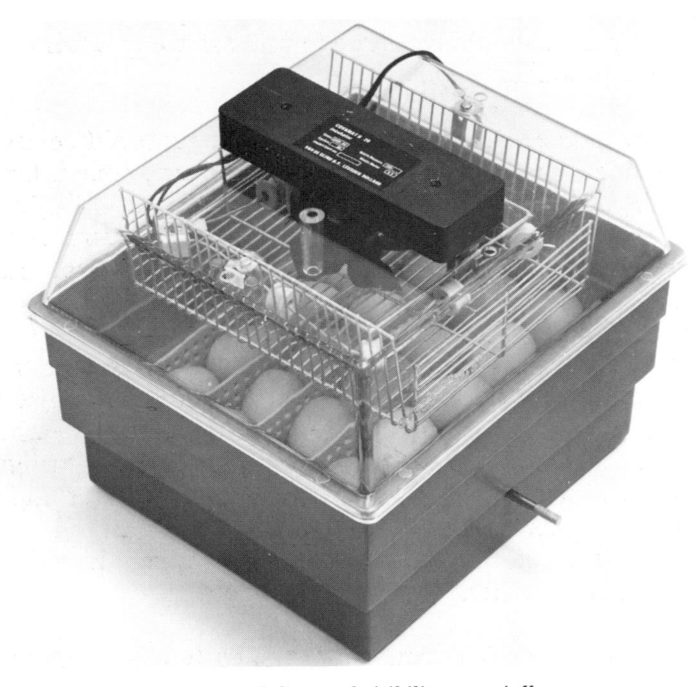

A 20-egg capacity incubator. It has good visibility, especially useful for educational purposes.

doubt be arrangements for classroom experiments of rearing to follow. However, thought should be given to what happens to the chicks hatched out in a school classroom. Any fertile eggs will serve the purpose of the demonstration, but it might be difficult to get rid of Heinz-type chicks when they are no longer tiny and fluffy.

Eggs

The fresher the egg on Day One of the 21 day incubation period, the better. Although fertile eggs will hatch after being stored for up to two weeks, the hatching percentage will be reduced.

Eggs have been designed to contain everything that the next generation of chickens will require during incubation and for the first 48 hours of life. Whether for eating or hatching, any extra care taken over their handling brings its own reward, either in extended shelf life or higher hatching percentages.

The following points should be borne in mind if the hatch is to be made from home produced eggs.

1. Collections should be made twice a day. This will protect them from variations in temperature and prevent them from being soiled. Once collected they should be stored at between 50 and 55°F (10 –13°C) with some humidity to reduce evaporation. (The amount of water lost in storage is significant enough to have caused problems in the EEC grading system. Producers were forced to install chilled storage to prevent eggs being eligible for sale at lower grades.)

Downward fluctuations in temperatures do not seem to matter as much as upward. Eggs have been known to hatch after being subjected to two or three days of very cold (though not freezing) weather. However, an egg which is raised to hatching temperature during a summer's day, will addle when the temperature drops at night. Strangely enough, or perhaps not, a broody will hatch out stale eggs more successfully than an incubator, provided the eggs have been stored by her and not for her.

2. Do not wash the eggs. It is better to set slightly dirty eggs than destroy the mucin protection by washing. If necessary, rub the shell gently with some sandpaper. The shell itself contains thousands of

minute pores so that the embryo can breathe. If the bloom is destroyed it is no longer a sterile environment. (Eggs for storage which are washed are more likely to take up any strong smell, especially fish, even when stored in the fridge.)

3. Never set cracked eggs. If the eggs are particularly precious or rare, a repair can be carried out by using nail varnish, or a cosmetic nail repairing product.

4. Use only the most perfectly shaped eggs possible. Do not set misshapen or very small pullet eggs. The average weight should be between 2 and 2½oz (56 –70g). Egg shape is an inherited trait, so a pullet hatched from a badly shaped egg is likely to lay a similar one. It is more likely that no chick will be hatched at all from a very badly shaped egg.

5. Prior to hatching, make sure the egg is stored correctly from another viewpoint. The air sac, so vital to the developing chick, is at the broad end of the egg and should be uppermost.

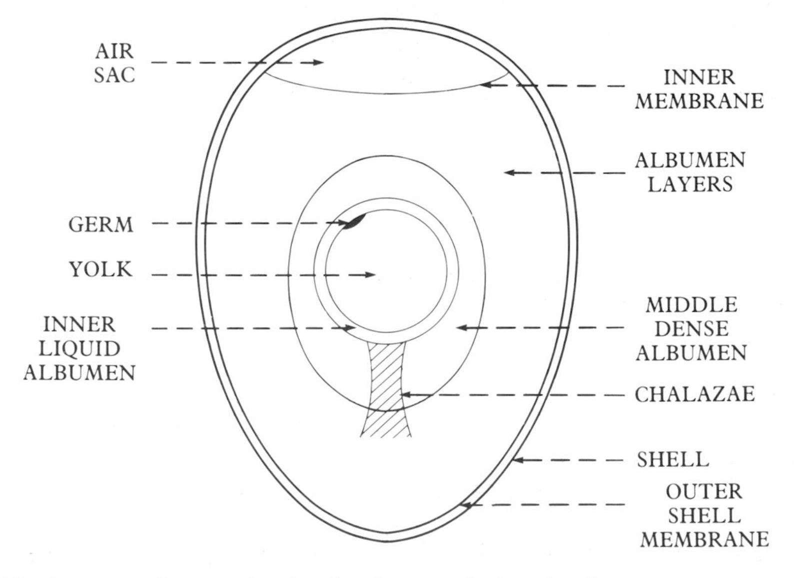

Fig 12 Structure of an egg showing the air sac at the broad end.

The 'Supaway' grader, which has a scale of all EEC egg grades and a scale of grams for weighing.

6. If collecting eggs over more than one day, mark each accordingly and note the results at hatching. (If eggs are to be stored for eating it is also necessary to store them by age, so that the supply is kept fresh. There is no need to store table eggs in a fridge.)

7. Make sure the eggs are fertilised. There is no secure way of telling before the hatch begins, other than making fertilisation possible. The

cockerel, who should be more than eight months old, must have been allowed time to settle in and about ten days to mate with the hens. For showing, a more mature cockerel with a proven record is likely to be used. No more than 10 to 15 hens should be run with a cockerel and 8 to 12 hens for a less mature cock.

One danger of buying in fertile eggs is that a lower percentage usually hatch, as they will be sold at a point in the season when the hens are laying prodigiously and are more likely to produce a higher proportion of infertile eggs.

There is no longer any need to enquire about blood testing for such egg transmitted diseases as Pullorum (B.W.D.) as this is now absent from Britain.

Incubation

The method of incubation isn't one of broody versus incubator. The choice must be made on what is most suitable for the circumstances under which they will be used. For example, it is not feasible to have a broody hen sitting in a school classroom for 21 days, and a 250 egg size incubator is not necessary for anyone with eggs from one show trio.

The main advantages and disadvantages of both methods are as follows.

INCUBATOR

Advantages

Eggs can be set at any time of the year
Good vision for what is going on and suitable for educational purposes
Will not desert the eggs
Easy candling which will not cause stress
No dirty nest to clean out

Disadvantages

Expensive to buy for small number of eggs
Brooding facilities required
Possible power cut (though some machines make provision for this)
Running and maintenance costs

Larger incubators are available with electronic control and automatic turning.

BROODY HEN

Advantages

Suitable for 12 eggs or less
No extra brooding equipment required
Daily turning not so crucial
The smell of a bad egg which is broken is less offensive under the hen
She can be used to brood chicks not hatched by her

Disadvantages

Only available at certain times of the year, and those dictated by the hen
Temperament. A broody may desert or destroy the eggs
Frequent handling is not always possible
She must be fed, watered, etc.

The comparative running costs are probably about the same. The cost of operating an incubator and subsequent heating for brooders has to be set against the cost of the hen, lost egg production and her maintenance and condition during the hatching period.

Some of the early arguments against what were then 'new fangled incubating machines' were that they tended to incubate the weaker members of the species, eventually undermining the hardiness and health of poultry. Whereas a broody hen would break a weak-shelled egg, or lose the weakest chick in the farmyard, the machines would do neither of those things.

These fears are groundless in regard to the eventual weakening of the species Gallus, but there is the point that nothing less than a perfect egg should be considered for hatching. A secondary guard is to use only those eggs laid by first class stock and a careful breeding programme.

A broody, in fact, gets away with a lot more than an incubator, especially in the matter of accumulated dirt in her nest. Although it should be kept clean, it isn't always possible to be too scrupulous at the risk of putting the hen off the nest, especially towards the nineteenth or twentieth day, when she might think she's been sitting there long enough.

TYPES OF INCUBATOR

Since all incubator instructions will vary slightly, there is little point in discussing the pros and cons of each one. However, there are two main categories of artificial incubating machines.

First, there are the electric cabinet types which force air round in a 'draught', and second, the still-air ones, which can be powered by oil, gas, electricity or paraffin.

Incubators can range in size from those suitable for a dozen or so eggs, right up to the commercial sizes carrying hundreds of eggs at one

Curfew offer a range of incubators with observation tops so the hatch can be watched.

The electric 'Agri' incubator has special insulation which maintains temperature for twelve hours after loss of current.

time. Some have a special feature of good vision, for educational purposes, and others specialise in good, automatic turning. (Some turn the eggs as many as six times a day.)

Brochures containing information about the newest range of incubators are available from the leading manufacturers, whose addresses can usually be found in poultry magazines. There is nothing really very complicated about any of them, and, as one manufacturer puts it – '. . . if all else fails, read the instructions'!

The older, second-hand or inherited incubator might be more of a problem. The most important thing is that the heat within the incubator does not raise the temperature of the growing embryo in excess of 100°F (37.8°C). Ideally, a temperature of 99.5–100°F (37.4 –37.8°C) should be maintained (exact temperature dependent on incubator type). This will need to be reduced towards the end of the hatch to allow for heat generated by the developing chicks. A normal hen's temperature is slightly higher than that of an incubator.

HUMIDITY

Humidity is also an essential factor in successful hatching and as a guide, a 60 per cent humidity should be maintained throughout the hatch. Some incubators have automatic turning trays, but otherwise each egg should be marked with a cross or an arrow and turned by hand three or four times daily, so that the embryo does not stick to the inside of the shell.

A broody will do this without thinking. If a paraffin operated incubator is being used, make sure there is sufficient fuel on hand to ensure there is no need to lose a batch of eggs through a drop in temperature just because the fuel ran out. Siting of an incubator is important, ensuring that there are no draughts. Exterior temperatures can affect those inside.

Candling

A process so called because the light of a candle was originally used to show up the degree of freshness of an egg. All eggs which passed through the packing stations in the 1950s had to be candled as they came from so many sources, and a degree of protection for the consumer was required. Candling consists of holding up the eggs, one at a time, in front of a strong light so that the contents become visible.

The most common use for candling now is during the incubation process, when it becomes necessary to know which eggs are fertile and which can be discarded. Usually on the seventh day and again on the fourteenth candling can take place.

A home-made device can be constructed by enclosing an electric light bulb in a box with a small hole cut in the opposite end to the light. The hole should be approximately 1¼ inches (3cm). A hand held candler is available from many suppliers and is of course very convenient.

The stages of the developing embryo can be followed by candling, though care should be taken to keep the egg on its side.

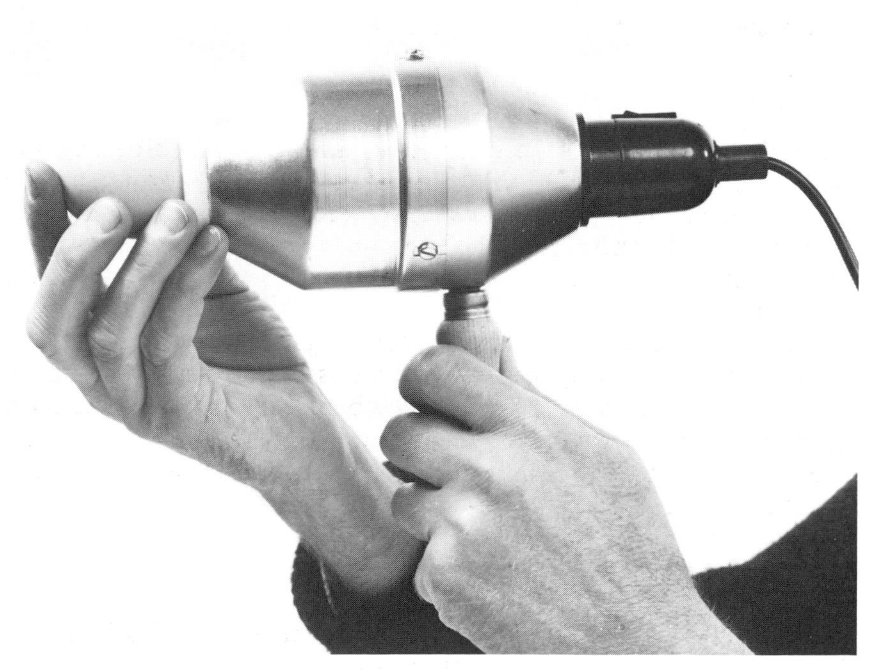

Eggs can be candled at the seventh and fourteenth days of incubation to detect infertile ones. Hand held candling lamps are convenient but expensive.

Day Number	Embryo Development
4	A small dark speck about the size of a pea should be visible. If the egg is clear right through it is infertile.
7	Blood vessels can be seen radiating from the pea sized embryo. The air sac can be defined as being paler than the rest of the egg. If an area of confused moving darkness is visible the egg has probably addled.
10	More blood vessels now visible, the air space clearly defined and the dark shape of the growing embryo should be moving.
14	The size of the embryo is now almost that of a 10p coin and movement can be seen. Blood vessels clearly defined.
18	About one third of the egg is clear and the rest filled with a black mass. Chick's beak can be visible.
19	The chick now almost complete and turning should cease.
21	Hatch.

Milestones in the 21-day incubation of a hen's egg.

A record should be kept of the various problems and successes during the period of incubation. It is important for future breeding programmes to know whether the fault lay with the mechanics of hatching, or whether there were a large number of infertile eggs. Infertility would be the result of either incorrect storage of eggs or a cockerel of dubious ability. It can also be caused by trying to breed from stock which has not reached sufficient maturity.

Natural Hatching

The first signs of broodiness in a hen are her reluctance to leave the nest box, the raising of her hackles together with a squawk and possible peck. Such a bird must be removed from the normal laying area. If she is not required for hatching, she can be put in a broody-breaker coop away from the other hens with her food supply cut down but with plenty of fresh water. She should then be marked before being returned to the flock. If she is going to be set on some eggs she can, if convenient, be left in situ until nightfall and then moved straight onto her new nest which need be nothing more than a good sturdy box, draught free and clean.

For the first day it is handy to have some china eggs available, to keep her in the mood while the real eggs are prepared or collected. It is sometimes possible to encourage broodiness by placing large numbers of china eggs in the ordinary nesting area.

At some stage the hen should be dusted with an insecticide. The hatching nest can also be treated for parasites before the nest bedding is put down. The latter can consist of a damp piece of turf, grass uppermost, to provide some humidity with a bedding of hay, straw, peat or even bracken on top. (The bracken if used should be dry and have had the hard stalks taken out.)

The box should be sited in a draught proof place but not necessarily in a shed, so long as it is protected from the elements, dogs or other intruders. Sometimes an ideal broody corner can be found but it proves difficult or impractical to fence it off. Provided reasonable supervision is possible, there is no reason why the hen should not be tethered during the hatch.

Once she is comfortable, with easy access to food and fresh water, there is little left to do until Day 21 (or Day 20 in the case of Bantam eggs). If only one candling is required, and it might be better not to disturb a good broody, it should be done at the tenth day. Any clears are then certain failures and can be removed from the nest box. If there are no signs that the hen has left the nest at all, she can be taken off and offered food and the opportunity to stretch her legs. In the second half of the sitting it is best not to disturb her, and certainly not after the eighteenth or nineteenth day. During the incubation time the hen will turn the eggs herself. If she has too many under her, all the eggs are in danger of being cooled in turn. A small Bantam can manage 8–10 of her own eggs while a large fowl could take a dozen.

Fresh food and water should be available at all times, although the hen will eat little during the first week. During the last six or seven days she might start to show an interest in the food (though not if it has sat there for the last 14 days) and wheat is generally a good ad lib offering.

The whole hatch should be completed within 24 hours of starting. Empty shells and any unhatched eggs should be removed, but not at the risk of disturbing mother and chicks. They should be left alone for as long as the hen is sitting comfortably. She is the best judge of when the hatch is complete and the first chicks will take no harm in her warmth. They need no food themselves for the first 36 hours, though the hen will. She has to provide the essential warmth and shelter for her brood.

If two or more broodies have been set simultaneously, and only one will be used for brooding, the chicks must be put together under one hen as soon as possible, bearing in mind the hen's accute eyesight. If a mixed batch of eggs have been set it wouldn't hurt to divide the chicks evenly between the hens by colour.

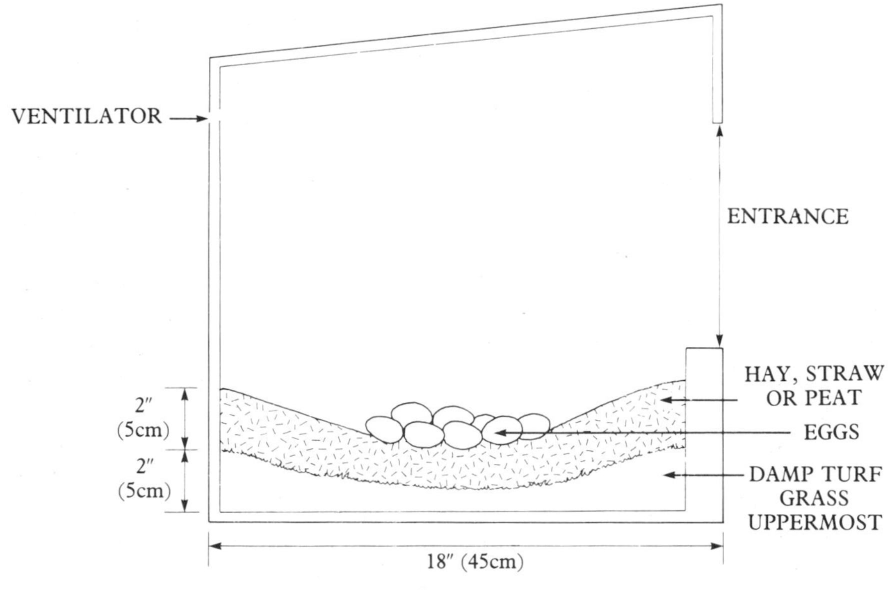

Fig 13 Hatching box suitable for broody hen.

It is possible to introduce incubator hatched day-olds to a sitting hen, provided it is done carefully. Use only a known hen, who has sat on china eggs for more than a week. During the late evening, introduce a single chick to her by placing it under her breast.

Leave her alone. After an hour or so it will be obvious from her quiet manner and clucking sounds that she has accepted it. The others can be gently added during the next half hour. There is no reason why she shouldn't accept them unless she is roughly handled or she hasn't been sitting long enough. Once all the chicks are under her, and there should be no more than 8 or 10 for a small Bantam, or 12 for a bigger hen, do not try to add to the number at a later date. Although hens cannot count, they do have the kind of eyesight which can tell black from white plus an uncanny knack of being able to tell day old chicks from three day old chicks. If a youngster from another brood gets mixed in with her own, and it is of a slightly different colour, it can be attacked and possibly killed.

Natural Brooding

Natural brooding means the transference of the hen and chicks to a small coop and run. The whole family will benefit from fresh air and access to the ground, and the hen will see that the chicks are not chilled. Protection from strong winds is a help. Unlike other young animals, chicks cannot stand low temperatures. They must be able to get warm as soon as they feel cold. This is sometimes decided by the mother hen, who fluffs up her feathers and clucks instructions to the infant chicks that it's time for a warm up.

Sometimes a coop with bars sufficiently close together to confine the hen but far enough part to allow freedom for the chicks is used. Although this is convenient, in that the hen cannot stray and is prevented from reducing the vegetable garden to dust, there is the disadvantage that she is not able to protect the chicks so easily from danger and predators. However, a run can be used and if the chicks only are actually on the grassy part they won't do half the damage that the hen will. This is a help if the family are put on a lawn or similar where they wouldn't normally be. It does help if the new chicks can be put on clean ground, but it is not always possible to fence such an area.

Commercially formulated chick crumbs are suitable feed and can be

A broody coop which can be closed at night helps to minimise predator risks.

Alternative all-in-one broody run.

offered ad lib. The chicks, especially Bantams, are often too small to use the larger water containers safely. A shallow tin with a large stone or brick in it will be adequate and a similar lid can be used for the food. Put only small quantities out at first (a handful twice a day for 10 –12 chicks is enough) as they will stand and scratch around in it while feeding. A Keyes tray is adequate and can be hygienically disposed of when soiled.

Artificial Brooding

If chicks are not going to be reared under a broody, some method is required for keeping them warm. Incubator hatched chicks, or bought in day-olds must be kept at a constant temperature of between 90 and 95°F (32–35°C) for the first week, decreasing by five degrees Fahrenheit (about 3°C) per week until the temperature is around 60°F (15.5°C) when the chicks are eight weeks old. Obviously, the temperatures will be affected by the prevailing climate and a minimum/maximum thermometer will help to record temperatures at nights. The whole brooding system should be on trial for 36 hours before the chicks arrive, in order that the litter should be warmed up and the draughts eliminated. If chilled early in life the chicks never really recover and will be stunted for the rest of their lives and not fulfil their potential.

By eight weeks of age the chicks can be weaned onto natural temperatures, care being taken for the first few nights. Management of the sixth and seventh weeks will depend very much on the time of year. If it is warm, the lamps could be switched off during the daytime as early as three weeks after hatching, provided careful monitoring of the thermometers is carried out.

The source of brooder heat can vary from a hay box or a single light bulb to purpose made equipment. There are the usual advantages of having the right tools for the job, and nearly all the incubator manufacturers also market brooder lamps, packaging the two at a slightly reduced rate. A typical lamp consists of a 250 watt infra-red dull emitter and a reflector, which provides all the required warmth on one unit of electricity for seven hours. Up to 75 chicks can be accommodated under one lamp. It is essential to carry at least one spare lamp, since something untoward will inevitably happen to the one in use after closing time on a Saturday evening.

A good home-made brooder can consist of an infra-red or dull emitter

The new gas brooders have a high degree of control and are especially useful in the later stages of growth.

The Maywick 'Titan' electric bright-emitter infra-red brooder can be used for up to 750 chicks.

Infra-red emitters for brooding come in a range of wattage and a small shade can spread heat for up to 75 chicks.

bulb suspended under an adjustable canopy. The surround is circular to prevent suffocation from huddling in corners and made from 2ft (60cm) high pieces of hardboard bolted together for easy dismantling and cleaning.

Overlapping sheets prevent draughts and allows the brooding area to be increased as the chicks grow. By moving the minimum/maximum thermometer around under the lamp it is easy to make sure there are no cold spots. This can also be assessed by the behaviour of the chicks themselves when they arrive. If they huddle together they are not warm enough, and if they lie constantly at the extremities of the brooding area they are too warm. If the temperatures are not correct the chicks will also cheep plaintively. A nice even distribution of chicks over the whole area is the aim.

An ideal litter for the brooding area is wood shavings. They are tightly packed in large bales, and a single bale lasts a very long time. Their size and composition make them more suitable than straw which isn't so cosy for the chicks unless chopped into small pieces. Long straw tends to 'pan' down and is not so conducive to scratching activities.

Greater emphasis used to be put on the benefits of outdoor brooding, heated by paraffin or gas. A particularly popular heater was the paraffin Putnam, which is still available for use in any home constructed brooding system.

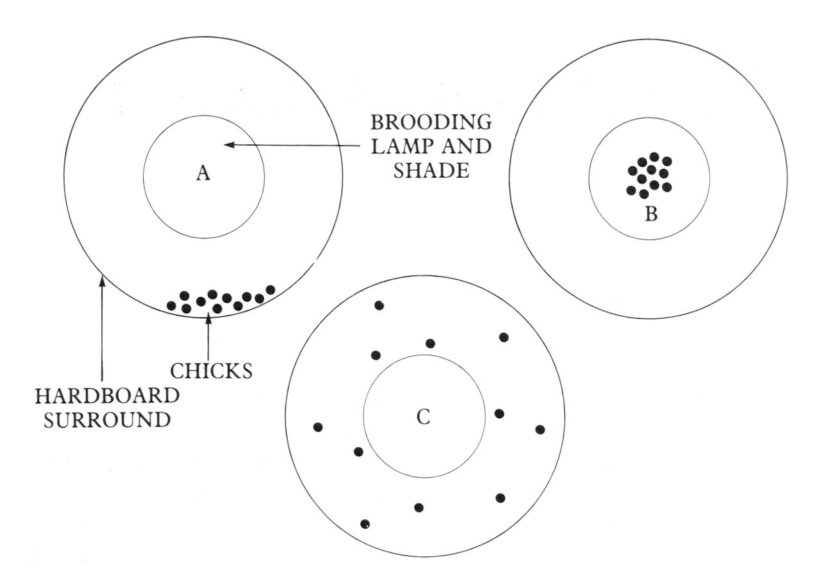

Fig 14 Brooding area A shows chicks too warm; area B shows chicks too cold; area C chicks at comfortable temperature.

Like natural brooding, the chicks reared outside do tend to feather-up more quickly and are said to be hardier. However, there are many dangers during the early weeks, the main one being the possibility of chilling. The heaters must be checked regularly to make sure they have enough fuel to last through a chilly night. Sufficient room must be left between the top of the heater and the flooring, not only to prevent 'spot' heating with associated dangers of huddling, but also to reduce the risk of fire. Chicks on range are also a prime target for predators. Everything from next door's cat to the local fox will be trying to find a way in.

Hay box brooding is often conducted out of doors in a warm season, and can work satisfactorily. Formerly called 'fireless brooding', the chicks are not generally put in the outdoor fold type of hay box until they are over two weeks. A similar shelf of hay could also be used in winter for extra warmth for older stock. A smaller, more enclosed type of hay box can be used as an emergency measure for very small chicks, the idea being to warm the chicks by their own heat retained by insulation. The major problem with this type of box is that after a day or so the nest becomes so fouled that it smells and becomes extremely unhealthy.

8 Health and Disease

PREVENTION

The prevention of illness or disease is preferable by far to its cure. Ill health is unprofitable, time consuming and heartbreaking, especially if it ends in the death of prize-winning stock.

The high standards of hygiene required to prevent some of the more common ailments are far better achieved by making them easy to adhere to. The rules of hygiene, for example, are easier to keep if feed troughs and water fountains are simple to clean. Before the huge strides made in the advancement of knowledge and control of disease, it was all very simple. Either the bird responded to one or other of the simple home remedies or it died. As one vet put it '. . . there were only three medicines in the cupboard, Karswood Poultry Spice, Stockholm Tar and Nicotine.' No more thought was given to hygiene in poultry than it was in the human context, and advice was passed on from farmer or fancier to the next, accurate or not.

Poultry diseases can, for ease of reference, be classified into two main groups. First, the specific diseases – which are those caused by a virus or bacteria, and which are infectious or contagious. Second, the non specific diseases – including nutritional deficiencies, housing or environmentally-caused problems, heredity or breeding difficulties and 'mechanical' disturbances, such as prolapse or impacted crop. Most of the specific diseases can be prevented by modern vaccines, but not all can be cured by antibiotics.

Most animals are seldom free from some type of parasite, or its effects, and this is particularly true with poultry, where individual treatment is not usually an economic proposition.

Nowadays, the prevention of disease and the preservation of health is considered in terms of flock management, and commercial farms could not operate without a high degree of prophylactic medicine. Many of the ailments caused by vitamin or mineral deficiencies are likely to be far less of a problem to the range or semi-range bird. The scientific knowledge about the nutritional requirements of poultry is such that the commercial compounds provide for all known requirements. As already mentioned in Chapter 6, pellets and mash will be date stamped with the shelf life of the vitamins and the medications stated alongside.

Generally speaking, the more space the poultry have to forage around in, the less chance there is of intensive flock diseases or vice. This does not mean that free range birds never get ill, but should any one bird be diseased there is less chance of the condition being spread rapidly. It should be appreciated that sometimes just six or a dozen hens kept in backyard style are, in effect, being kept in semi-intensive conditions and are as prone to vice and disease as commercial flocks. Range birds are also at risk from pests and diseases that intensively housed chickens have never heard of, and are guaranteed a large parasitic worm intake.

However, the advantage that non-intensive birds have is their hardiness, and higher degree of resistance to disease. Some pure breeds are said to be more resistant than others to particular disease while some non-specific diseases are caused by breed idiosyncracies. For example,

Easily cleaned plastic drinkers help to prevent the spread of disease.

'hard feathered' game birds are far more likely to suffer from infestation of external parasites than looser feathered breeds.

There are well over 100 identifiable poultry diseases, and to name and describe them all is unlikely to be helpful to anyone keeping hens on a small scale. Many of them occur but rarely and would probably need to be identified by veterinary investigation and confirmed by post mortem. Many of the symptoms of disease are common to one or more condition, which again makes identification and diagnosis difficult.

However, the novice, or near-novice, will want to know something of the one or two well known and more common diseases. Even commercial producers only concern themselves with some of the lesser known problems if and when they occur on their farms.

The most dangerous time for the chickens, and the time when they are most likely to fall victim to disease, is during the first six weeks of life. From the start, the chick must struggle to maintain its health against the incessant attack from cold, hunger, parasites, virus, bacteria, predators, and others. It therefore needs all the help it can get, and winning the fight can be greatly assisted by cleanliness and informed management. The greater the individual resistance, the less chance germs have of gaining the upper hand.

It is well known in all circles of husbandry that livestock in general always do better on a 'clean' site, i.e. one which has not previously or recently been used by the type of stock being carried. While it is not always possible to continually graze or fold the hens onto new ground, it is possible to go some way towards breaking the lifecycle of intestinal worms, thus attempting to prevent the 'tired ground' factor. Young chicks in particular should never be run on the same ground as the main flock. The same applies to housing and runs. If the management practice is to buy in POL pullets, and replace them after one year, the accommodation should be thoroughly cleaned as soon after they are cleared out as possible. It should then be rested before the new batch arrives.

Not only does the litter and ground carry infection, but so also do the stock themselves. Mixing birds of different ages should be avoided if at all possible, especially for the first few weeks after new stock arrive. 'Imported' stock should be isolated and acclimatised. The most likely birds to be brought in regularly are show birds, or a breeding trio. Isolation is a requirement for all such imports. Most fanciers also isolate their show stock after it has, of necessity, come into contact with other show entrants. There are plenty of rules about the health of show stock, but no guarantees.

Any second-hand equipment, especially incubators or brooders, should be thoroughly cleaned and disinfected. It is no use just chucking a few pails of disinfectant over it and hoping for the best. Any muck or dirt has to come off first as that is probably where any disease is, if it's there at all. Really thorough cleansing more or less does the job of a disinfectant, but it doesn't work the other way round.

Veterinary attention to single birds is only justified if it is a particularly valuable bird, or there is suspicion that an ill bird shows symptoms of a Notifiable Disease. That means contacting the Divisional Office of the Ministry of Agriculture. The most obvious disease to come under this heading is fowl pest, also called Newcastle disease, and not to be confused with fowl pox which has different symptoms and is not so serious as fowl pest.

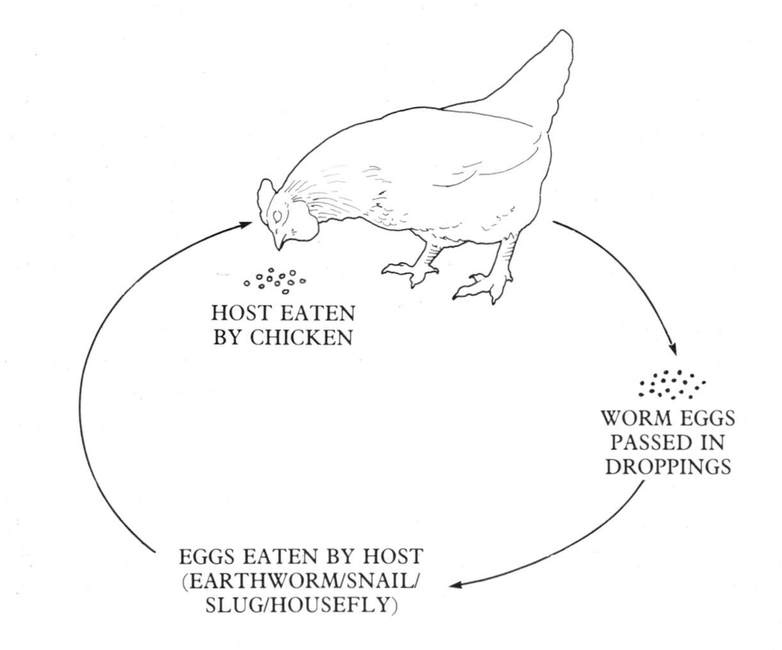

Fig 15 The cycle of worm infestation which must be broken.

Fowl Pest

In older veterinary books, fowl pest is also referred to as the plague, and Newcastle disease as pseudo-fowl pest. The confusion over names came about after a very severe outbreak of the plague occurred in Doyle near Newcastle-on-Tyne in 1926. The then identified disease was referred to as Newcastles, but did not become notifiable until a further outbreak in Britain in 1933 when the Fowl Pest Order was introduced. Fowl plague is exceedingly rare in Great Britain, but the virus is so similar to Newcastle disease that only laboratory investigation can differentiate between the two.

Egg Eating

Although the small scale poultry keeper is not usually troubled by vice associated with intensive production methods, there is one which can be a problem, and one which has been eliminated by battery cages. Egg eating can break out at any time, in any place, and once it has begun in a flock, no matter how small, it is extremely difficult to get rid of.

It usually begins accidentally, when a hen breaks an egg in the nest box, either because there are one or two eggs already there when she gets in to lay, or the hen herself lays a soft-shelled egg which breaks when laid or soon afterwards. The first hen to visit the box is almost bound to investigate the broken egg, and egg eating has begun. The hens will soon learn to break the shells open themselves and will be copied by other hens in the flock.

If ever good management was the key to prevention, it is here. Darkened, clean nest boxes are a must, and frequent egg collection will not only reduce the number of eggs lying around in the boxes, but any broken ones can be quickly removed, the boxes thoroughly cleaned and new litter put in.

If egg eating does occur, it is easy to spot during the early stages, but less so later on when the hens get professional at it. Then it will be a question of trying to account for reduced production, and since there is often no evidence of broken shell in the nests, it will be difficult to track down.

It is possible to 'blow' an ordinary egg and fill it with curry powder, mustard and peppercorns which have been soaked in paraffin. This decoy egg is placed in the nest and, when pecked at by the bird, should be unpleasant enough to discourage it from further shell breaking. Unfortunately, it does not very often work and all the other eggs smell of paraffin. It could be worth a try using just the curry powder and mustard, but it is generally a waste of time.

Pests

The degree of pest problems will depend very much on the type of housing. In deep litter buildings, and laying sheds, the areas round the water drinkers will be those worst affected.

There is little that can be done about the variety of small beetles and moths which will colonise during the time the poultry is resident, but the ordinary housefly can become a problem, not just in the poultry accommodation but also in the farmhouse kitchen.

A female fly can lay 500 eggs per month, so an ideal fly breeding establishment is not very conducive to good neighbourship in an urban area. Although flies breed throughout the year, the problem is worse in summer, when warm, damp conditions occur more frequently. During the warmest times it will be impossible to prevent any flies from thriving, but if they are controlled at source the problem should not get out of hand. Insecticide blocks hung from the ceiling will help, taking advantage of the habits of the housefly in alighting onto hanging objects. In some larger rearing houses it would be possible to hang a length of string with smaller lengths hanging from it, from one end of the house to the other, having soaked the whole thing in insecticide. Naturally, care must be taken with poisonous substances and with selection of the correct insecticide. It is necessary to control flies, as they are a major risk in spreading infection. They, and other insect pests, can also be a cause of stress to the fowls.

Worms

Another problem more likely to be encountered by range birds than intensively housed ones is that of worms. Not the sort they spend hours scratching around to find, but the parasitic types which sap the health of the stock which in turn renders it more liable to other infection and causes unthriftiness.

Worms will also be a problem in overcrowded housed stock and where land is fowl-sick, i.e. ground which has been used for chickens continuously for months or years.

The two main groups are round worms and tapeworms. Normally, parasitic worms are not a problem for producers of table birds, as the birds don't live long enough to become badly affected.

ROUND WORM

The large round worm can be as long as 2½ inches (6.35cm), whitish grey in colour and affects chicks from two months of age onwards. If left untreated, appetite is affected and emaciation is progressive. If large numbers of the parasites become mature and the condition left without treatment, the birds will die.

TAPEWORM

Tapeworms may be present in strings of up to 10 inches (25cm) long and will cause unthriftiness and a drop in egg production. An old remedy for tape worms is a mixture of turpentine in olive oil (½–1cc of turpentine in 10cc of oil for a large bird).

Chemical treatment is available for all types of worm. Symptoms of worm infestation are apparent when birds stand around in a depressed state, sometimes attempting to swallow continuously. (Though birds do this sometimes in very hot weather as well.) Birds which cough or retch, and are losing condition, almost certainly have a worm problem.

GAPE WORM

In chicks it could be the gape worm which inhabits the windpipe of the birds, and this is best prevented rather than cured. All young stock should be put only onto clean ground. It has been found that the earthworm can carry the gape worm larvae for at least four years.

If any doubt arises as to diagnosis of worm infestation, the symptoms should be outlined to a vet, who can recommend a suitable vermicide. A similar prescription could be used prophylactically, on advice from the surgery. This need not involve an expensive visit from the vet to the stock. A single bird can be taken to the surgery. Usually, if just one bird is ill, or has died (whether from worms or anything else), it is not worth the expense of a visit. However, if several birds are affected by illness, it is advisable to seek professional advice.

External Parasites

One of the easiest health risks to control is that of external parasites. Lice, fleas, mites and bugs of varying types are all sitting around just waiting for the opportunity to live on good, healthy chickens.

There are eight different kinds of lice, and all live on the feathers or skin of the bird. They lay their eggs at the base of the feathers and after fourteen days are ready to begin chewing at the feathers and skin scales of their hosts. Different lice like different parts of the body. The neck louse, for example, causes damage to the neck and head areas.

HEN FLEA

The hen flea can be found in open ground as well as in enclosed houses. After attacking the birds, they then seek refuge in dirty corners of houses, nest boxes and around food dispensers where they lay their eggs. Fleas in nest boxes can be particularly disturbing to the laying hen and could be an indirect cause of broken eggs. The recommended dusting with flea powder for sitting hens is due to the restlessness caused by attacking fleas.

BED BUG

If a hen owner comes out of the chicken house scratching and itchy, it is likely that the bed bug is present. Like the fleas, they do not always live on the host. After engorging themselves with blood, they go off to a safe hide-out to digest their feed and lay their eggs. The bed bug is not peculiar to poultry and can live as happily on other animals, humans amongst them.

RED MITE AND NORTHERN FOWL MITE

The red mite is not, as might be suggested, red, but a pale grey. Only after feeding from the bird's blood does it become bright red. Unfortunately, the red mite can stand not being fed for some time, and can therefore be difficult to get rid of from housing merely by resting it. It feeds at night and can be responsible for causing anaemia in the fowls. The red mite does not need to remain on the bird, but the second most common mite, the northern fowl mite does, and subsequently is more affected by insecticides.

SCALY LEG MITE

The third most common mite is the scaly leg mite and, as its name suggests, it confines its activities to the leg scales, causing swelling and chalky deposits. This is very uncomfortable for the bird and evidence of its existence is more obvious than with other mites. This is an extremely common mite and a most persistent one. The bird's legs should be scrubbed with a solution of *benzyl benzoate*. An old toothbrush is very effective. The birds should be treated every ten days. Not only is treatment of the birds essential, but eradication of the mite itself from the perches and surrounding woodwork is also necessary. If the condition is not diagnosed and treated, it will cause the birds great distress and they will begin to peck at their legs and feet in an attempt to relieve the irritation.

TREATMENT

Apart from reducing the number of places where external parasites can breed, treatment of lice, flea and mite infestation is by application of an insecticide to the bird itself. This is not always appreciated by the patient, and to help prevent stress and excitement to everyone concerned in the operation, the birds should be held upside down, that is head down.

One handler and one assistant is ideal. If there is no help available, some control of the bird can be gained by restricting its ability to run away, perhaps by tying its legs together. Once the bird has a quick rush of blood to the head, and has tired itself with wing flapping, the initial frantic protest should exhaust the bird sufficiently for it to be docile while being treated. If help is not available for the application of the chemical, the bird can be temporarily 'hung' which will leave the handler with both hands free. Such handling is rather too rough for a broody, who anyway is likely to be easier to handle. A more convenient method is application of insecticide by aerosol, and one favoured by exhibitors.

Range birds use a dust bath to rid themselves of external parasites, but this should be considered a behavioural practice and not relied upon to do a complete job of pest removal. Derris powder, however, can be added to a dust bath if it is an official one and regularly used by all the stock.

Diagnosis

Diagnosis of most poultry disorders is usually difficult and bewildering, not only to the beginner, but to the practised hand. First of all there has to be an indication that all is not well. This is usually seen by observation that one or other bird is not behaving as normal. The appearance of a bird can indicate whether it is suffering from a specific or non specific disease. If the comb shows signs of deterioration, discoloration or pigment change, but her appetite is still good, there is a good chance that the fault lies in the diet, and there is probably a deficiency somewhere.

Any loss of appetite, change in gait or general depression would indicate something more serious which must be diagnosed and dealt with. In layers, a falling off of eggs is not always one of the first symptoms of ill health. As long as a hen is taking in sufficient protein on which to produce eggs, the reproductive system is not always the first area affected.

Immediate reaction, on spotting anything amiss with one or more birds, is to isolate the patient and watch carefully for other symptoms upon which to base a diagnosis. Watch the rest of the stock as carefully to see whether another bird is developing a similar condition.

Although a small scale poultry keeper or breeder can go for years without encountering any major problems, is it useful to know about some other specific and non specific diseases which could affect his or her stock.

B.W.D (Bacillary White Diarrhoea) or Pullorum Disease

An acute infectious disease primarily affecting baby chicks, but can also occur in adult fowl. Due to years of blood testing and an eradication policy on the part of the Ministry of Agriculture, B.W.D. is now almost completely absent from Britain. It used to be necessary to check that parent stock of pullets or chicks were free of B.W.D. but this is now no longer as important as it once was.

B.W.D. is caused by *S. Pullorum* and belongs to the Salmonellas group of bacteria. It is spread by 'carrier' birds and their eggs. A single egg could infect a whole batch of incubating eggs. (A good reason for thoroughly disinfecting second-hand brooding and hatching equipment). The disease, being blood transmitted could be discovered by blood testing, which all commercial breeders and hatcheries did, and carriers culled.

Coccidiosis

This affects chicks up to six or seven weeks of age with symptoms becoming obvious at about four or five weeks. Death is caused by haemorrhaging which happens very soon after any blood in the droppings will have been observed.

It is caused by various species of coccidia, which not only affects young chicks but other young animals as well. To some degree, coccidiosis can be considered endemic, from the point of view that it is usually present where poultry is kept, and if the chick is exposed to only a mild attack it will become resistant.

Prevention is the best cure here. A coccidiostat is usually incorporated in commercially prepared chick crumbs. However, clean, dry and fresh litter, together with all the usual good hygiene and good husbandry practices, is as good a preventative as anything else. The levels of medication are kept down so that natural immunity is allowed to develop while some protection is given. Unfortunately, the health of the chick is irreparably damaged before signs of coccidiosis are apparent, but if it is diagnosed correctly, some treatment for those not affected can be recommended by a vet.

Mareks Disease

This is caused by a virus, and affects the central nervous system. Chicks in the first few weeks of life are most at risk, but symptoms sometimes do not manifest themselves for several months. All commercial hatcheries vaccinate automatically for Mareks between hatching and despatching. In some hatcheries, hand vaccination has been replaced by automatic machines, which can vaccinate up to 2,000 chicks an hour. Although vaccination does not guarantee freedom from an outbreak, it considerably reduces the risk. It is as well to check from the supplier that the stock is protected before it is brought onto the holding.

Impacted Crop

Occasionally, a chicken's crop will become impacted with grassy or fibrous material. If a blockage occurs in the gullet it prevents the food from travelling normally from the crop to the gizzard. Usually the cause is due to the bird having eaten too large a quantity of coarse material, poor quality grass, hay, straw or other litter. However, if more than one bird shows signs of a pendulous crop, further investigations should be made.

Care must be taken to prevent such impacting taking place when young chicks are being transferred from brooding conditions to grass range, when they are not used to the new diet. Also, there is a tendency for winter housed stock to eat straw, in the absence of anything else to do. Sometimes the situation can be sorted out by holding the bird upside down and massaging the crop. If possible, try to get some olive oil into the crop area first.

Egg Binding

If a hen is seen repeatedly entering and leaving the nest without laying her egg she has probably got a very large or misshapen egg stuck in the oviduct. (Or she is a persistent egg eater and she is always on the look-out to see whether any more have been laid.) Unless treated, the constant straining will result in a prolapse. Massage, and the introduction of oil into the vent area will help. The egg can be broken to assist in its removal.

Prolapse

Often the result of a hen constantly straining to remove an obstruction in the oviduct, such as an over-large egg, or broken one, or as a result of very heavy egg production. The reddish tissue can be seen protruding from the vent area, but if the condition is not severe it can easily be missed, as the tissues will return to their proper place after an egg is laid. However, other hens will not miss it and a nasty case of cannibalism can occur. Culling is the only cure.

Bumblefoot

This condition is brought about by continual damage to the sole of the bird's foot, usually as a result of heavy birds having to jump down from perches set too high. Certain bacteria invade the damaged tissues, and set up infections which swell the pad of the foot. This is another one of those conditions which are better prevented than treated, but should signs of discomfort be evident in a bird's gait, the feet should be washed carefully in warm, salty water. If necessary, apply medication (penicillin cream if possible) and bind with padding and a thin cotton tie-on bandage. Do not use sticky plaster.

A condition similar to bumblefoot can occur in stock which are housed on deep litter. If the litter is not kept dry and clean, a hard mass of droppings and straw can accumulate on the pad of the birds' feet. If the birds go outside for any part of the day they usually dislodge the worst before it builds up, but in permanently housed flocks it will only get more impacted. Each bird must be treated individually, its feet being soaked in warm water. Do not on any account try to prise off the balled matter while dry. Once it has soaked through it can be scraped off. Once the birds are treated, it does no harm to dip their feet in a mild disinfectant or salt water solution. The wet and dirty litter needs of course to be removed.

Crooked Toes

Sometimes a young chick will have one or two toes turning sideways instead of being straight. This is not a disease as such, and more often than not does not affect the performance of the bird.

For some reason the increase of crooked toes in poultry is associated with artificial rearing, but so far no particular reason has been identified and proved. It can happen with naturally reared chicks as well, but not so regularly. It is possible that it is an inherited trait, and therefore any hen with such a deformity should not be used for breeding. It could also be caused by the rapid growth of artificially brooded chicks.

The condition is distinct from curled toe paralysis, which is due to a vitamin B_2 deficiency. There is no treatment for crooked toes. If, however, young chicks show signs of curly toe paralysis, the toes of the affected ones will turn inwards and curl underneath the feet. Obviously they will become less able to walk. If the condition is very advanced when it is first noticed (and this might happen in a large brood of young chicks), it is as well to cull the affected chicks straightaway, rather than waste more food on producing a bird which will not be able to carry out a productive or comfortable life. If caught in time, the remedy is to examine the quality of the feed. It is possible that, if the parents of the chicks were deficient in the vitamin, there may have been insufficient passed into the hatching egg and thus the chick started life with a deficiency.

Deficiency Diseases

Very few range or semi range birds are likely to suffer from vitamin or mineral deficiencies in adult life, but care must be taken to ensure that artificially reared chicks have a balanced diet. All commercial chick crumbs will have the required levels of vitamins present, with their shelf life stamped on the bag label. However, if large amounts of food are left lying around under the intense light and heat of a brooding lamp, its nutritional value will deteriorate. The main vitamin requirements are:

Vitamin A – to prevent rickets
Vitamin B Complex – to prevent curly toe disease in chicks and decreased egg production in hens and markedly low hatchability
Vitamin E – to prevent nervous disturbances and unstable gait.

Folk Remedies

There are one or two remedies and tonics which have been handed down through generations of small scale poultry keepers.

EPSOM SALTS

Epsom salts (magnesium sulphate) is a useful medicine for several poultry ailments. Although a saline purge, and therefore useful when hens appear to suffer from constipation, it is given when loose droppings indicate internal disorders. The salts are best given in the drinking water and for flock treatment, one tablespoon is sufficient for each pint. For individual doses, one quarter of a teaspoon in a little warm water can be administered directly to the bird by means of a pen-filler or glass dropper.

CIDER VINEGAR

Cider vinegar is also useful in maintaining good health, and is reputed to assist growing chickens to feather out quickly and ensure tasty lean meat in broilers. The liquid can be given in the drinking water; only a capful per pint is needed.

GARLIC

Garlic has a history as a deterrent against internal worm infestation, and a small amount of powdered garlic can be sprinkled on feed once a week. Whole, fresh garlic can be put into the drinking water, either loose or tied in a piece of muslin with a heavy stone. The use of garlic is unlikely to taint the eggs unless given in huge quantities. Garlic was commercialised during the 1920s and marketed under the trade name of Yadil, and according to the company's Veterinary Book would cure all known poultry diseases, specific or non-specific. Garlic, like other herbal remedies, should be thought of more as a preventative medicine than as a cure. Rue is said to have similar properties to garlic when it comes to deterring parasitic worms.

POULTRY SPICE

Poultry spice is still marketed more or less as it has been for many years, though its use is now primarily for winter-housed stock which do not have daily outside access. It is used by Fanciers who are aiming to bring their stock up to peak condition for a show following an enforced moult.

Ingredients vary slightly according to the supplier, but it consists mainly of the essential minerals and vitamins required for good health.

COD LIVER OIL

Cod liver oil is a most versatile and useful aid to health, especially during winter. Its high levels of vitamins A and D make it a valuable guard against deficiency diseases. A teaspoon daily is enough for a dozen small chicks, and during the worst of the winter it can be given at the rate of 2oz (56g) to every 6½lbs (3kg) of mash.

STOCKHOLM TAR

Stockholm tar is very messy but can be useful if feather pecking breaks out, or when a single bird is attacked by another for whatever reason. Some of the tar smeared on a bare patch will deter other stock from pecking it.

SEAWEED

Seaweed can be fed to hens, and is said to be most beneficial to their general health. It can be dried and sprinkled onto the daily ration, or hung up when fresh for the birds to peck at. Care should be taken that only small quantities are available ad lib as the high salt content can cause other problems, as mentioned in Chapter 6.

9 The Fancy

It is a natural progression for anyone interested in keeping chickens for eggs, to wonder about the possibilities of specialising in a pure or rare breed, and as a result become interested in the Fancy.

Exhibiting poultry has always attracted innumerable small scale breeders, either to increase the value of their stock or just for the joy and challenge of showing. It has also attracted Royal patronage since 1843 when the young Queen Victoria acquired the famous *Cochins*. The Royal example did much to set the Fancy on its way to becoming an established part of agricultural shows, and then to independent exhibitions run exclusively by the Fancy for the Fancy.

When the railway network opened up, it not only threw up the possibility of new markets for commercially produced eggs, but the opportunity for wider travel made it possible for fanciers to journey to shows. The 1840 Penny Post improved communications generally, and contributed to the establishment and backing of the poultry sections of the large agricultural shows.

Royal support for poultry breeding has continued. Queen Alexandra specialised in several varieties of Bantam, while George V had poultry at Windsor. The Royal interest continues today, with H.M. Queen Elizabeth the Queen Mother's *Buff Orpingtons*, kept at Sandringham in Norfolk, giving the lead to fanciers in the Eastern Counties. H.M. The Queen was Patron of The Poultry Club in her Silver Jubilee year 1977, and The Queen Mother is its current Patron.

The Poultry Club maintains breed standards and promotes a public awareness of its declared standards as well as publicising the worth of today's British poultry stock. It organises the National Championship Show each year and there are affiliated shows throughout the country, where special prizes are awarded to members. A regional show scheme is operated at selected larger agricultural shows during the year.

Other national and regional shows are organised by The Rare Poultry Society, which was formed in 1969, and has nearly 100 rare poultry breeds under its protection, ranging from *The Booted (Sabelpoots) True Bantams* to the *Transylvanian Naked Neck*.

ENTERING THE FANCY

But what of the newcomer, who, having mastered the basic husbandry skills now wishes to widen his or her interest and develop it into an absorbing hobby?

It is unlikely at this stage that the novice has any stock worth taking to a show at all, and it is best that visits to shows should be in the role of spectator, prepared to get the 'feel' of showing. The show day activities are often the culmination of years of planning and careful breeding by the exhibitors. Watching the actual judging of both birds and eggs will give an opportunity to acquire experience about what to look for when training show birds. An enquiry to the Steward will usually be rewarded by a contact in the local Club and an introduction to an exhibitor. Fanciers are usually more than ready to welcome a newcomer and to share their own experiences. Apart from winning, the pleasure for many fanciers is the chance to socialise with other enthusiasts.

Due to strict livestock import restrictions, exhibiting overseas is limited, but many fanciers attend shows in Europe and North America to compare breeds and further their knowledge.

Starting Up

Having taken a good look at the many different breeds and varieties, a decision to specialise in one or two only is the best advice which can be given. Producing quality stock isn't achieved overnight. A few birds bundled into a crate, and crossed fingers, just won't do. The experience gained from competing is better achieved when there is some hope of being a winner. Not that anyone does it for the financial rewards, as prize money is often only a few pounds.

The summer shows are the best start for a new fancier. Most agricultural shows have a poultry section and, although these shows coincide with an end-of-season condition, and thus it is more difficult to present decent stock, the competition is less fierce.

The majority of exhibitors keep Bantams, often from the several advantages of their size, but also because their popularity has ensured that there is usually good competition and a better chance of a good market for eggs from winning birds.

Aim

The aim of breeders is to produce stock, Bantams or Large Fowl, which are as close to the Standard as possible. The Standard provides the opportunity for common comparison and consists of 100 points per bird.

Not all features receive the same number of points. Each breed is judged according to the British Poultry Standards which have allocated a certain number of points to breed features. For example, the comb and lobes of the *Minorca* amount to more than all the head points of, say the *North Holland Blue*, whose merits lie in other directions.

It is accepted among poultry fanciers that 'type makes the breed and colour the variety'. This means that type refers to the general characteristics, shape and bearing of a breed, while the colour of their plumage decides which varieties of that breed they are. Such differences in colour, shades thereof, and where they appear on the body are all refinements which distinguish one variety from another.

Clubs

Most breeds have their own Club, or Society, and once decisions have been made about the breed which appeals most to the newcomer, it is as well to contact that Club and get as much information as possible about the breed and the Club's activities. Those breeds which are either rare or extinct are looked after by The Rare Breed Society, and all others by The Poultry Club. The Club Secretary can advise on a reputable source of potential winning stock. Either young stock, discarded for one reason or another by the breeder, or a breeding trio, are both more suitable than hatching eggs. Beware, however, of some less selective vendors who can be seen buying unknown stock from markets and re-selling them to the unsuspecting as pure breeds.

Housing and Caging

Some special housing and caging facilities are required for anyone thinking of breeding for show. Training cages are necessary, and should be the same size as those used for showing. All the necessary equipment for bathing, training and general management can be kept together in one room, or garden shed, in which can also be stored travelling cases. The penning room, or shed, is where the essential training and pre-show handling is done. This handling is best continued throughout the chicken's life and it is a vital factor in successful exhibiting. The younger the bird when it is first handled the better. No

Efficient use of space can make a good penning room in limited areas.

Careful handling has taught this Game bird not to be afraid.

judge wishes to be confronted by a nervous bird flapping madly around covering all and sundry with sawdust.

Birds need to be used to penning. When they are first put into the cages they are sometimes understandably frightened. They need to overcome this fear and be calmed down well before show day. Some fanciers prefer a concentrated training session immediately prior to the show, while others practise year-round sessions. Once a bird has been trained it is easier to handle the second time.

After becoming used to penning, they then have to be unafraid of the judge, whose hand coming at them unexpectedly might unnerve them. A good ploy here is to make sure their food is given by hand so that they gradually get used to the cage door being opened and shut.

Sometimes a judge will use a stick, and this can be very alarming to an untrained bird. Have a judging stick handy in the penning room and use it occasionally to gently move the bird around the pen. Watch a judge use his stick at a show and use the same movements in training.

Chicks which are handled when very young are not afraid of being approached. Getting them to accept titbits from the hand – cheese is ideal for this – is good for show handling. Pick them up several times a day, and sometimes from their perches at night, and they will grow up very tame and not intimidated by human activity. Some breeds need special handling to show them at their best in the ring – *Moderns* for example must stretch tall. They should be tempted with a titbit to reach up, while *Pekins* must do the opposite and can be floor fed. *Indian Game* are encouraged to stand 'wide' and need plenty of exercise to develop their muscles. Very showy birds must learn to sit quietly in the palm of the hand to be examined, and the better behaved the bird the more chance it has of capturing the judge's approval – provided it also has the correct conformation and feather condition.

PREPARATION

The object of getting the birds to their peak of condition will not be achieved a week or even a month before the date, unless the competitor has a wide choice of stock. The decision about which bird is to be shown should be made when the application for a place is made. The last minute sprucing up is done the day before, but the body and plumage condition is the result of good, consistent management throughout the year. Last minute entries to shows are not usually allowed, and entrance fees to most shows need to be paid well in advance of the event. If a definite choice of stock is made then, work can begin to make sure the birds are at their best on the day.

Diet

All fanciers will have their favourite diets for year-round health, and for special meals prior to showing. While in the training pens, a morning feed of wheat or maize can be followed by greenstuffs at noon, and a good spiced mash in the evening. Water and grit must be ad lib and fresh. Care must be taken to ensure that the drinker cannot easily be knocked over, as wet wood shavings and droppings will combine to undermine any bathing done close to show day. If maize is being fed, its use should be restricted to dark plumed birds as its colour can affect the lighter coloured varieties. Tinned cat and dog food is a favourite titbit, as well as bread and milk with a few drops of cod liver oil added. Canada and hemp seed ate traditionally given pre-show.

It is best if these feeds are supervised and not left in the pens. Apart from the advantage of human presence at regular intervals in the training pens, stale food will be discarded and insufficient rations will be taken. The eagerness (or not) of the bird for the next feed will indicate that all is well with its appetite and so, probably, all is well health-wise. Appetite is a good indication of health. If a bird has been force moulted, to bring it into condition for a winter show, some poultry spice is often added to the food as an extra precaution and an aid to building up protection against disease. No bird is immune from an attack by bacteria or virus infection, and the fact that home conditions are clean and disease free will not guarantee that other competitors' are. The regular show people are always aware of hygiene, but occasionally an enthusiastic newcomer can be a weak link in the armour.

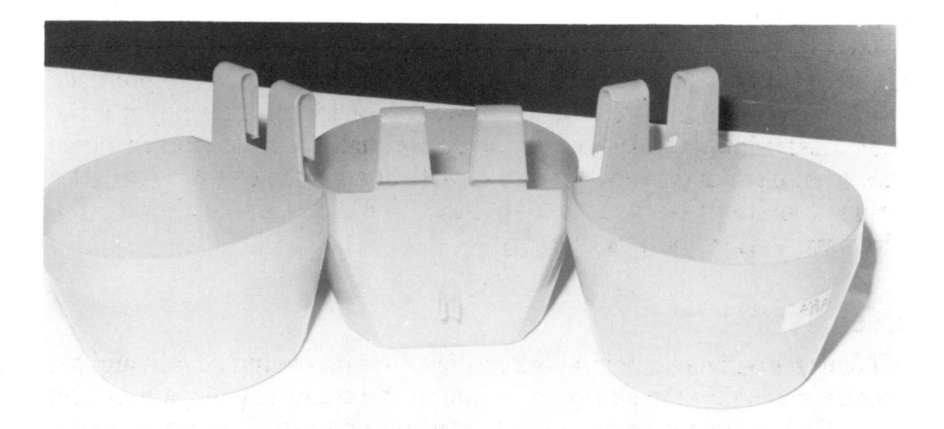

Small plastic drinkers are useful for fixing on to a cage wire to prevent spillage.

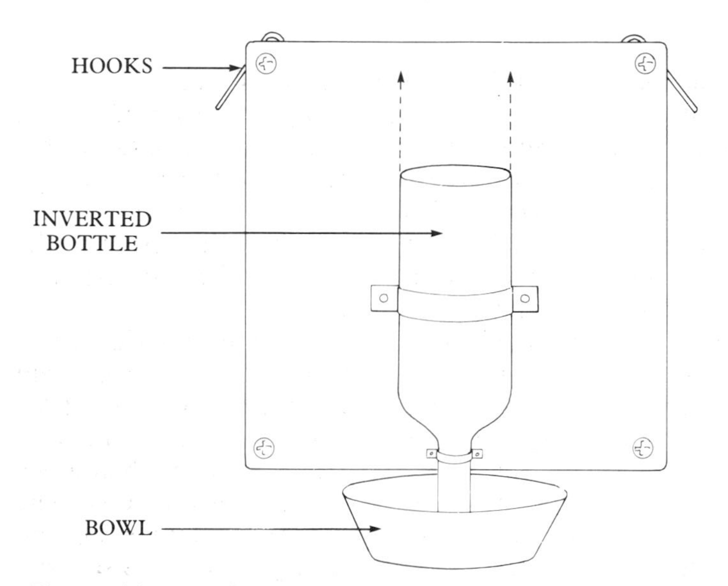

Fig 16 A home-made water dispenser fixed to the side of the traning pen will prevent spillage.

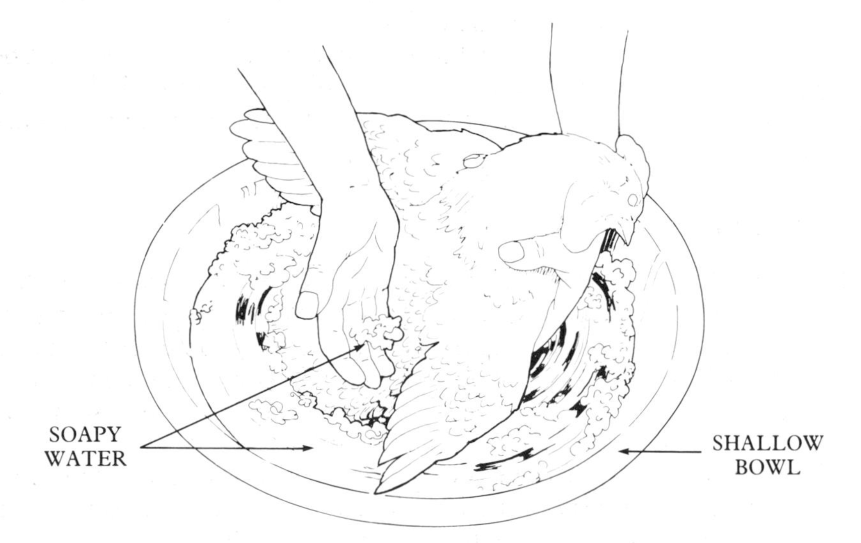

Fig 17 Show birds should be washed gently in a shallow bowl using ordinary soap flakes.

Washing

Although no amount of washing and preening will improve a poor quality bird sufficiently to win in decent company, any bird can be improved by being produced in an immaculately clean condition. No matter how good the bird, it will not receive the highest awards if dirty.

Most breeds, including 'hard feathered' ones can be washed. Preferably, in the case of *Old English Game*, two or three days prior to the show. A shallow open basin is best, as the bird can be held with one hand and washed and rinsed with the other.

Dirt should be removed carefully from beneath the leg scales with a plastic tooth pick or similar, after legs have been given a good scrub with a nail brush. Ordinary shampoo or soapflakes can be used for washing the feathers, using the same method as washing hair, and making sure that all soap is removed by rinsing. An ordinary hair dryer will complete the job and prevent damage to the feathers which could be caused by towelling. Too good a dry can cause brittle feathers. Drying should stop before the plumage is completely dry. The bird itself will preen the feathers when he or she is returned to the peace of the training pen. Dark plumaged birds do not need washing all over, though the head, legs and feet always benefit from such attention.

Having spruced up the show birds it follows that their litter must be kept scrupulously clean and dry. The training pens should always be washed down and aired between use and, if there has been any disease trouble, disinfected as well.

Transportation

The birds should be transported to the shows in strong boxes or wicker baskets. Custom made carriers are available, but other modes of transport are acceptable, so long as escape en route is not possible, air movement is as free as possible and the bird is not cramped. Boxes should be marked with the owner's name and address. There should be a lining of some sort inside to prevent damage to feathers.

The birds should not travel in the same box as the rest of the show gear. A bird which has spent a long journey jumping around to avoid rolling aerosols and bottles will not appear at its best.

Transport can come in a variety of shapes and sizes.

Competitors will be sent their show tags well in advance of show day.

Game birds, resting between shows.

At the Show

On arrival at the show, after finding the allotted cages, it is a good idea to clean round the pen before unpacking the birds. Often there is a lot of dust, or even wetness on the partitions. Then the bird itself should be cleaned, to remove any straw or droppings which have adhered to the feet in transit. Most fanciers have a concoction of oil and astringent with which they wipe over the bird's head. Usually olive oil is used and mixed with vinegar or wine spirits. There are rules which prohibit what is called 'tampering' so it is as well to read them. Most shows are held in accordance with the rules of The Poultry Club.

Labels will have been sent to all exhibitors, and stock usually has to be at the showground the day before the opening. Smaller shows will have their own regulations about timings, and these should be checked out well in advance. Usually the larger shows take responsibility for the stock while it is on the showground and stewards are appointed to deal with any problems. The showground Association will also have a veterinary inspector who will be called if there are any stock deaths or suspicious symptoms of illness.

SHOWING EGGS

As well as exhibiting birds, some fanciers take part in showing eggs. There are authorised standards for internal and external judging of eggs which have recently been reviewed. As with the stock, points are awarded for shape, size, colour and appearance. The various aspects are awarded to a total of 100 points.

Maran, *Welsummer* and *Barnevelder* are three breeds which are particularly prized for their dark brown egg colour. This is at its best when the bird first comes into lay. Only the most perfectly shaped eggs should be offered for exhibition, and no polishing or colour additives are allowed. No egg which is double yolked or stale will be judged. Freshness is assessed by the height of the yolk. The higher the yolk the fresher the egg, an important 35 out of 100 points worth of freshness. Eggs should always be judged for their contents on plates, as saucers have the disadvantage of the yolk height disappearing into the recess.

Eggs come in varying shades of brown, white and mottled and there are different classes for each. There is a 'tinted' class, which covers numerous shades of almost any colour, though some shows specify exact shades while others will accept just about anything. The blue, green or olive eggs of the *Araucana* breed usually come under this heading, or the 'Any Other Colour' class. Bantam eggs must not weigh more than 1½oz (42.5g).

Glossary

Addled A fertile egg where the embryo has begun to develop but died before completion.

AOC Any other colour.

AOV Any other variety.

Auto Sexing Method by which chicks can be sexed at hatching by their down colouring (*see* Chapter 7).

Axial Feather The small feather separating primary and secondary wing feathers.

Bantam Mini version of the larger breeds, being approximately one quarter the size.

Barring Feather markings of equal size and alternating stripes, as fine as possible, of two definite clear colours across feathers.

Battery The housing for intensive egg production birds. Statutory regulations are laid down regarding minimum sized cages and the whole system is under EEC review.

Blastoderm The germ in a fertile egg, identified as a spot in the yolk.

Blood Spot The spots of blood formed on the yolk of an egg, which are caused by the rupture of small blood vessels in the egg follicle. Probably an inherited weakness.

Bloom The gloss and good condition of egg shells. Also applies to plumage and leg scales of the bird.

Glossary

Broiler Birds especially bred in huge quantities for the meat market. Sold as male, female or 'as hatched' with a potentially good food conversion and quick finishing facility.

Brooder Appliance for artificial brooding of chicks. Sometimes referred to as *Hover*.

Broody Female bird who ceases to lay and shows willingness to sit on eggs and rear offspring.

Candling Means of determining whether hatching eggs are developing by placing the egg in front of a strong light. Also used to detect blood spots. A clear egg denotes freshness. In incubation a clear egg after the fourteenth day indicates infertility.

Cape Feathers Those feathers between the neck and shoulders.

Capon Formerly a cockerel which had had its male characteristics changed by hormone implantation (at about five months of age) or surgical operation (at about eight weeks of age). The need for caponisation has been superseded by broiler breeding, where male birds reach maturity much quicker without the need for a caponising practice.

Chalazea The thick 'cords' of albumen on each side of the egg yolk, holding it in place.

China Eggs China, or pottery, eggs are 'pretend' eggs used to test broodiness and possibly save the waste of real hatching eggs should a broody refuse to sit.

Clean Free from disease, not newly arrived from the laundry!

Clear An egg, which when candled at the seventh day of incubation, shows no sign of embryo development.

Clipped Wing *See* Chapter 3.

Cockerel A male bird of less than 12 months.

Contract Rearer Professional rearer of commercial day old chicks, layers and broilers, subject to a particular company's requirements.

Crop A sac at the base of the throat where food is ground before passing on to the gizzard and through the digestive system. Lactic acid renders the food suitable for digestion.

Cuckoo Barring Irregular feather barring, where two colours run into each other rather than being distinct.

Culling Disposal of birds which are either unfit for egg production or have ceased to lay economically. A breeder culls his flock in order to produce only the best in future generations. Culling can also take place when sexing the young chicks in hatcheries.

Cuticle *See* Mucin.

Day Old Chick up to 48 hours old or until it has its first feed, which is likely to be after its journey from hatchery to purchaser.

Dead in Shell Refers to chicks which have died inside the shell, unable to break out, either partially or completely.

Debeaking Removal of the tip of the top half of the beak, carried out only by commercial companies, and is done to discourage the vices of feather pecking or cannibalism of intensively housed birds.

Deep Litter A system of housing poultry which allows the litter to remain in situ and break down over a period of two years.

Down The initial soft, hairy covering of baby chicks, sometimes referred to as 'fluff '.

Dressed Poultry Carcasses which have been prepared for sale by being plucked, gutted and sometimes packaged.

Dual Purpose Breed of bird suitable for both egg production and meat, but excels at neither.

Dubbing Removal of the comb, wattles and ear lobes, leaving a smooth head. Not a practice of commercial poultry production but of fanciers.

Dust Bath A bath of fine dust, cinders or dry earth which the bird propels up to her skin between the feathers to help remove body parasites.

Endemic Disease which is constantly present but only flares up when conditions are right for its manifestation.

First Cross The progeny of a mating between two pure but different breeds, varieties or strains.

Flight Feathers Alternative name for the primaries – the ten long feathers hidden when the wing is closed normally. Referred to as 'flights'.

Free Range Complete freedom for the birds to forage and choose where to rest, eat, bath and lay their eggs.

Frizzle Plumage giving the impression of being brushed the wrong way! The curled feathers should turn backwards towards the bird's head. The *Frizzle* breed is usually a Bantam, and is bred for exhibition.

Hackles The long, narrow feathers extending down the neck of a fowl. Also the saddle plumage of the male, which are pointed.

Hard Feather Show term to denote close, tight feathers, such as those found on game birds.

Hardening Off The gradual weaning of young stock from artificial, constant temperatures at brooding to lower ones and finally natural air temperature.

Heavy Breed Any breed in which the female hen averages more than 5½lb (2.48kg) at maturity.

Hen Female bird over one year old.

Hover *See* brooder.

Hybrid Birds produced by scientific breeding to give the result of progeny capable of specific production performances.

In-breeding The mating of closely related stock.

Incubator An enclosure capable of controlled environment in which fertile eggs are placed for the required time for hatching. Can be fuelled by electricity, gas, paraffin or oil.

Keyes Tray Standard papier maché egg trays, moulded for tier storage.

Lay Away The habit of range birds to lay their eggs in places other than the official nest box, usually inaccessible places.

Light Breed Opposite to Heavy breed.

Mandibles The horny upper and lower parts of the beak.

Mash Feeding stuff comprising cereals compounded to provide the necessary balanced diet. Fed either in its dry, powdery state or mixed with water.

Meat Spots in Eggs Nothing to do with meat! Caused by concentrations of calcium getting mixed up with the albumen formation. Usually an inherited weakness.

Mucin Protective covering of egg shell, acts as a barrier to water and bacteria. Destroyed by washing. Also called the *cuticle*.

Notifiable Disease If the symptoms of a notifiable disease are suspected, notably fowl pest, there is a statutory obligation to report it to the Ministry of Agriculture's Divisional Office.

Out Crossing Introduction of fresh blood by the use of a male from a different strain to the one being used.

Pencilling Small markings or stripes across the feathers, for example, *Pencilled Hamburghs.*

Point of Lay (POL) Pullets which are approaching the time when they will lay their first egg.

Pot Eggs *See* china eggs.

Preening Process of grooming in which the bird rubs its beak against oil glands situated at the roots of the tail feathers. The secreted oil is used by the bird as a dressing for its plumage.

Primaries Section of wing flight feathers, hidden when wing is closed.

Pullet Young female bird of 12 months or less.

Pure Breed Birds of either sex which have the blood of only one breed in them and are genetically stable. They breed 'true' amongst themselves and when used with another breed produce a known hybrid.

Saddle The rear part of the back in a male bird.The same area in a female is called the cushion.

Secondaries The inner group of shorter feathers in the bird's wing, which are used in flight. These large feathers are visible when the wing is closed.

Sheen The glossy surface on black plumage. Called lustre in all other colours.

Soft Feather Show term indicating breeds other than game birds.

Spurs The horny spike on the inside of the shanks of male birds and some females.

Standard Specifications and judging points of breeds as laid down by The Poultry Club.

Strain A particular family of birds with no outside blood introduced. A breeder can develop his own strain or strains of any breed.

Table Eggs Old fashioned term to describe those eggs produced for eating as opposed to hatching.

Taint Eggs, because of the porous nature of their shells, can acquire any strong odour which surrounds them. Feeding fish meal is supposed to cause taint in eggs, but it is more likely due to the eggs coming into contact with droppings in the vent area.

Trio One male and two females, usually purchased together for specialised breeding.

Variety A recognised branch of any breed, usually known by distinctive colouring or marking. There is usually more than one variety of each breed.

Veterinary Investigation (VI) Centres Laboratories under the authority of the Ministry of Agriculture, which carry out post mortem examinations and otherwise investigate disease and disease problems.

Waterglass A chemical preparation used for preserving a glut of eggs for winter use. Not so much used now with facilities for year-round egg production, and the fact that eggs do not always maintain their fresh taste for quite so long as some manufacturers state.

Watery Whites Often found in eggs laid by birds coming towards the end of a long laying season. Can also indicate respiratory problems. A cloudy or milky-like white is often caused by an excess of riboflavin (vitamin B_2).

Wattles The fleshy appendages on either side of the base of the beak, more developed in male birds.

Useful Addresses

Ministry of Agriculture, Fisheries & Food (MAFF)
Whitehall Place, London SW1A 2HH.

MAFF (Publications)
London SE99 7TP.

Gleadthorpe Experimental Husbandry Farm
Meden Vale, Mansfield, Nottinghamshire NG20 9PF.

The Poultry Club of Great Britain
General Secretary: Mike Clarke, 30 Grosvenor Road, Frampton, Boston, Lincolnshire PE20 1DB.

Rare Poultry Society
General Secretary: R.J. Billson, Alexandra Cottage, 8 St Thomas's Road, Great Glenn, Leicestershire LE8 0EG.

The Soil Association
86–88 Colston Street, Bristol, Avon BS1 5BB.

The British Egg Information Service
Bury House, 126–8 Cromwell Road, Kensington, London SW7 4ET.

British Poultry Breeders Association
High Holborn House, 52/54 High Holborn, London WC1V 6SX.

British Poultry Association
Albany House, 324–6 Regent Street, London W1R 5AA.

Chicken's Lib
P.O. Box 2, Holmfirth, Huddersfield HD7 1QT.

Index

Page references for illustrations are indicated by italic type.

Index